RAND NATIONAL SECURITY RESEARCH DIVISION

AUSTRALIA'S NAVAL SHIPBUILDING ENTERPRISE

Preparing for the 21st Century

JOHN BIRKLER
JOHN F. SCHANK
MARK V. ARENA
EDWARD G. KEATING
JOEL B. PREDD
JAMES BLACK
IRINA DANESCU
DAN JENKINS
JAMES G. KALLIMANI
GORDON T. LEE
ROGER LOUGH
ROBERT MURPHY
DAVID NICHOLLS
GIACOMO PERSI PAOLI
DEBORAH PEETZ
BRIAN PERKINSON
JERRY M. SOLLINGER
SHANE TIERNEY
OBAID YOUNOSSI

For more information on this publication, visit www.rand.org/t/RR1093

Library of Congress Cataloging-in-Publication Data is available for this publication.
ISBN: 978-0-8330-9029-4

Published by the RAND Corporation, Santa Monica, Calif.
© Copyright 2015 RAND Corporation
RAND® is a registered trademark.

Limited Print and Electronic Distribution Rights

This document and trademark(s) contained herein are protected by law. This representation of RAND intellectual property is provided for noncommercial use only. Unauthorized posting of this publication online is prohibited. Permission is given to duplicate this document for personal use only, as long as it is unaltered and complete. Permission is required from RAND to reproduce, or reuse in another form, any of its research documents for commercial use. For information on reprint and linking permissions, please visit www.rand.org/pubs/permissions.html.

The RAND Corporation is a research organization that develops solutions to public policy challenges to help make communities throughout the world safer and more secure, healthier and more prosperous. RAND is nonprofit, nonpartisan, and committed to the public interest.

RAND's publications do not necessarily reflect the opinions of its research clients and sponsors.

Support RAND
Make a tax-deductible charitable contribution at
www.rand.org/giving/contribute

www.rand.org

Preface

The Australian government will produce a new Defence White Paper in 2015 that will outline Australia's strategic defense objectives and how those objectives will be achieved. It will consider future force structure options for the Australian Defence Force that align strategy with capability and resources. Part of the process of producing the Defence White Paper is to examine an enterprise-level shipbuilding plan that brings together navy capability requirements, available resources, and the future composition of the Australian shipbuilding and ship repair industrial bases. To inform this process, it was necessary to conduct our analysis in parallel with the ongoing development of the Force Structure Review. The resulting demand profiles used in this report were therefore used as exemplars as the government considers its final force structure requirements through the White Paper process.

Historically, Australia has acquired ships from overseas (e.g., the *Charles F. Adams* guided missile destroyer and the first four *Oliver Hazard Perry* guided missile frigates); more recently, the focus has shifted to acquiring ship designs from overseas and modifying those designs to meet Australian requirements. All or parts of these ships have then been built in Australia. Typically, ship support activities for needed repairs and modernization have been accomplished in-country by Australian public- or private-sector organizations. However, demands for Australia's shipbuilding industrial base have been sporadic over the past 20 years, and the various peaks and troughs in demand have led to a decline in skill resources and an inefficient industrial base. Problems with the shipbuilding industrial base have been reflected in the recent difficulties experienced with constructing the new air warfare destroyer and have raised issues concerning the cost of building new ships in Australia versus having the ships built in a foreign shipyard.

A major question facing the nation's government is whether Australia should support a naval shipbuilding industry or buy ships from foreign shipbuilders. This is a complex matter with many facets and subsets. The issues often center on cost trade-offs, but there are also important national sovereign and strategic concerns. There should also be synergy between shipbuilding and the support of those ships once they enter service, although typically, shipyards tend to focus on either building new ships or supporting in-service ships. What is important both in building and in supporting ships is knowledge of the design of the ship so that construction, maintenance, and modernization can be conducted in a cost-effective manner. If the answer to the basic question is a desire for an Australian shipbuilding industry, subsequent questions involve what future demands are needed to permit the industrial base to operate effectively and efficiently and how the industrial base assets should be organized.

RAND Support of the Australian Department of Defence

In September 2014, RAND was engaged by the Australian Department of Defence to undertake a series of materiel studies and analysis activities. The Defence White Paper team tasked RAND to inform the development of an enterprise-level plan for naval shipbuilding for consideration by the government. To develop this plan, the authors focused on three important tasks:

- Provide an understanding of the current Australian shipbuilding capability and gauge how alternative acquisition strategies might affect both the capacity of the industrial base and the total cost of the enterprise.
- Compare the costs of Australia's naval shipbuilding industry with overseas manufacturers that produce platforms of comparable size and scope.

- Assess the economic costs and benefits of government investments in Australia's naval shipbuilding industrial base under the various enterprise options.

Between September and December 2014, RAND produced a series of interim studies to present insight into its ongoing analysis, to inform the research sponsor of early findings, and to provide a means to elicit feedback as the work continued.

This report is intended for an audience that has some familiarity with naval shipbuilding. Comments or questions on this report should be addressed to either of the project leaders, John Birkler (email: birkler@rand.org; telephone: +1-310-463-1924) or John Schank (email: schank@rand.org; telephone: +1-703-413-1100, extension 5304).

Management Framework

This research was sponsored by the Australian Department of Defence and conducted within the Acquisition and Technology Policy Center of the RAND National Security Research Division (NSRD). NSRD conducts research and analysis on defense and national security topics for the U.S. and allied defense, foreign policy, homeland security, and intelligence communities and foundations and other nongovernmental organizations that support defense and national security analysis. For more information on the Acquisition and Technology Policy Center, see http://www.rand.org/nsrd/ndri/centers/atp.html or contact the director (contact information is provided on the web page). For more general questions about RAND operations in Australia, please contact our RAND Australia director at moroney@rand.org.

Contents

Preface iii
Figures xi
Tables xvii
Summary xxi
Acknowledgments xli
Abbreviations xliii

CHAPTER ONE
Introduction 1
RAND Research Objective 2
Structure of the Report 4

CHAPTER TWO
Australia's Naval Shipbuilding and Ship Repair Industrial Bases 5
What Makes Naval Surface Ship Production and Sustainment Unique 6
Australia's Naval Shipbuilding Industrial Base 8
   Shipbuilding Workforce 11
Ship Support Industrial Base 13
   Differences Between Shipbuilding and In-Service Ship Support 13
   Composition and Capabilities of the Current In-Service Ship
      Industrial Base 16
   How the Fleet Places Demands on the Support Industrial Base 20
Future Demands for Support 21
   Ability to Meet Future Demands for Ship Repair 24

Observations on the Shipbuilding and Ship Support Industrial Bases 25
Australia's Unique Market Niches: Warship and Submarine
  Production and Support 27

CHAPTER THREE
**Australian Department of Defence's Planned and Projected
  Warship Acquisitions** 31
Current Shipbuilding Programs 31
  Landing Helicopter Dock Project 31
  Air Warfare Destroyer Project 32
Future Shipbuilding Programs 33
  Future Submarine Project 33
  Future Frigate Project 33
  Supply Ship Project 34
  Mine Hunter and Hydrographic Ship Replacements 34
  Pacific Patrol Boats 35
Emerging Short-Term and Longer-Term Demand Gaps 36
Alternative Strategies for Australian Naval Ship Acquisitions ... 38
  Modify an Existing Design to Be Built in Australia 39
  Modify an Existing Design to Be Partially Built and Outfitted in
    Australia 41
  Buy a MOTS Ship from Overseas and Use It Without
    Modifications 42
Australianization Issues for National Ship Programs 43
  Design and Build Issues 44
  Relationship Between Shipyard and Designer 45
  Observations About Adopting Foreign Designs for Australia 45

CHAPTER FOUR
**Using Indigenous Australian Industry to Address the Short-Term
  and Longer-Term Gaps in Naval Ship Demand** 49
Analytical Approach, Model, and Assumptions 50
Addressing the Short Term: Baseline Analysis 53
Sustain a Fully Capable Australian Shipbuilding Industrial Base . 61
  Addressing the Short Term 61
  Addressing the Longer Term 77

Structure of a Fully Capable Australian Shipbuilding Industrial
    Base 84
Sustain a Limited Capability Australian Shipbuilding Industrial
    Base 87
Addressing the Short Term 87
Addressing the Longer Term 92
Structure of a Limited Capability Australian Shipbuilding
    Industrial Base 93
Sustain Only the Australian In-Service Ship Support Industrial Base 94
Summary 94

CHAPTER FIVE
**Benchmarking Australia's Naval Shipbuilding Industry with
    Comparable Overseas Producers** 99
Cost Benchmarking 101
    Approach 101
    Caveats 102
    Input Benchmarking 103
    Comparative Benchmarking 109
    Parametric Analysis 121
    Influence of Exchange Rate 122
    Summary of Cost Benchmarking 123
Schedule Benchmarking 124
    Approach 124
    Caveats 126
    Source of Data and Initial Analysis 126
    Weight as a Proxy for Complexity 130
    Summary of Schedule Benchmarking 130
Observations on Australian Shipbuilding Cost and Schedule 131

CHAPTER SIX
**Examining Economic Pros and Cons of Australian
    Government Investments in Various Naval Shipbuilding
    Enterprise Options** 133
Economic Multipliers and Their Uncertain Implications 133
Newport News Shipbuilding Case Study 134

Austal USA Shipbuilding Case Study 135
Gripen Case Study 135
Discussion 136

CHAPTER SEVEN
Conclusions and Recommendations 139
Detailed Findings 139
  What Are the Comparative Costs Associated with Alternative
    Shipbuilding Paths? 139
  Is It Possible for Australia's Naval Shipbuilding Industrial Base to
    Achieve a Continuous Build Strategy, and How Would Such
    a Strategy's Costs Compare with the Current and Alternative
    Shipbuilding Paths? 145
  How Do the Costs of Acquiring Vessels Domestically Compare
    with Acquiring Comparator(s) from Shipbuilders Overseas? 146
  How Much Do Expenditures Connected with Warship Building,
    Maintenance, and Sustainment Add to Australia's Economy? 147
Summary Implications 148

APPENDIXES
**A. Shipbuilding in Australia: A Brief History and Current
    Shipyard Production Facilities** 151
B. Shipbuilding Model and Assumptions 165
C. Sensitivity Analysis 191
D. Exploring the Option of Producing Offshore Patrol Vessels 213
E. Survey of Australian Shipbuilders and Ship Repair Industries ... 227

Bibliography 243

Figures

S.1.	Workforce Profile for Building Air Warfare Destroyers and Future Frigates (Base Case)	xxviii
S.2.	Total Labor Costs of Base Case and Alternative Shipbuilding Construction Paths	xxx
S.3.	Total Schedule Delay of Base Case and Alternative Shipbuilding Construction Paths	xxxi
S.4.	Ships Used in Comparison Benchmarking	xxxiv
S.5.	Average Keel to Commission Schedule	xxxvi
2.1.	Major Australian Shipyards and Their Current Roles in Naval Shipbuilding and Repair	10
2.2.	Comparison of Production Skills Demanded by Representative Australian Shipbuilding Versus Ship Maintenance Projects	16
2.3.	Percent Distribution of *Anzac* Ship Alliance Labor Costs for Recent Maintenance Projects	17
2.4.	Number of Surface Combatants in Maintenance, 2014–2019	21
2.5.	Average Age of Royal Australian Navy Fleet, 2014–2046	24
4.1.	High-Level Architecture of RAND's Shipbuilding and Force Structure Analysis Tool	51
4.2.	Workforce Profile for Building Air Warfare Destroyers and Future Frigates (Base Case)	54
4.3.	Workforce Profile for Lessening the Short-Term Production Gap (Base Case)	55
4.4.	Unproductive Man-Hours, by Workforce Sustainment Level (Base Case)	57

4.5.	Total Labor Costs, by Workforce Sustainment Level (Base Case)	57
4.6.	Schedule Implications of Sustaining 5 Percent of Peak Workforce Demand (Base Case)	59
4.7.	Schedule Implications of Sustaining 20 Percent of Peak Workforce Demand (Base Case)	60
4.8.	Total Labor Costs and Schedule Delays, by Workforce Sustainment Level (Base Case)	60
4.9.	Workforce Profile for Building Future Frigates Starting in 2018	62
4.10.	Schedule Implications of Building Future Frigates Starting in 2018	63
4.11.	Total Labor Costs and Schedule Delays, by Future Frigate Construction Start Date (Full Capability Path)	64
4.12.	Workforce Profile for Adding a Fourth Air Warfare Destroyer	65
4.13.	Schedule Implications of Adding a Fourth Air Warfare Destroyer	66
4.14.	Workforce Profile for Building Patrol Boats at Major Shipyards Starting in 2020	67
4.15.	Workforce Profile for Building Patrol Boats at Major Shipyards Starting in 2017	68
4.16.	Schedule Implications of Building Patrol Boats at Major Shipyards Starting in 2017	69
4.17.	Workforce Profile for Building Four Offshore Patrol Vessels at Major Shipyards	70
4.18.	Total Labor Costs of Building Three, Four, or Five Offshore Patrol Vessels at Major Shipyards	71
4.19.	Total Schedule Delay Relative to *Anzac*-Class Retirements of Building Three, Four, or Five Offshore Patrol Vessels at Major Shipyards	72
4.20.	Workforce Profile for Scenario 1, Longer-Term (Full Capability Path)	77
4.21.	Workforce Profile for a Continuous Build of the Last Six Future Frigates with a Drumbeat of Two	80
4.22.	Future Frigate Force Structure with Drumbeat of Two, 2026–2041	82

4.23.	Total Labor Costs for Building Three, Four, or Five Offshore Patrol Vessels in the Short Term, with a Future Frigate Drumbeat of Two	83
4.24.	Total Schedule Delay for Building Three, Four, or Five Offshore Patrol Vessels in the Short Term, with a Future Frigate Drumbeat of Two	84
4.25.	Workforce Profile for Building Air Warfare Destroyers and Future Frigates (Limited Capability Path)	88
4.26.	Total Labor Costs, by Workforce Sustainment Level (Limited Capability Path)	89
4.27.	Average Cost per Full-Time-Equivalent Worker, by Workforce Sustainment Level (Limited Capability Path)	90
4.28.	Schedule Implications of Sustaining 5 Percent of Peak Workforce Demand (Limited Capability Path)	91
5.1.	Relative Oil, Chemical, and Gas Plant Construction Costs	107
5.2.	Relative Construction Costs, Based on First Marine International Shipbuilding Productivity	108
5.3.	Exchange Rates for Australian Dollar to U.S. Dollar and Euro over the Past Five Years	124
5.4.	Time-Series Plot of Keel-to-Commission Schedules	128
5.5.	Average Keel-to-Commission Schedule	129
5.6.	First Ship's Metric Tons of Full Load Displacement Versus Months from Keel to Commissioning	130
7.1.	Total Labor Costs of Base Case and Alternative Shipbuilding Construction Paths	141
7.2.	Total Schedule Delay of Base Case and Alternative Shipbuilding Construction Paths	142
A.1.	Ship Production Phases	163
B.1.	Workforce Profile for First-of-Class Future Frigate (Full Capability Path)	176
B.2.	Workforce Profile for First Follow-On Future Frigate (Full Capability Path)	177
B.3.	Workforce Profile for One Outfitting-Only Future Frigate (Limited Capability Path)	178
B.4.	Workforce Profile for One Patrol Boat	179
B.5.	Workforce Profile for One Littoral Multirole Vessel	180

B.6.	Workforce Profile for Existing Air Warfare Destroyer Program	180
B.7.	Workforce Profile for a Fourth Air Warfare Destroyer	181
B.8.	Workforce Profile for One Offshore Patrol Vessel	182
C.1.	Effect of Level of Effort on Total Labor Cost (Full Capability Path)	194
C.2.	Effect of Level of Effort on Total Schedule Delay (Full Capability Path)	195
C.3.	Effect of Drumbeat on Total Labor Cost (Full Capability Path)	196
C.4.	Effect of Drumbeat on Total Schedule Delay (Full Capability Path)	197
C.5.	Effect of Learning on Total Labor Cost (Full Capability Path)	198
C.6.	Effect of Learning on Total Schedule Delay (Full Capability Path)	198
C.7.	Effect of Hiring Rate on Total Labor Cost (Full Capability Path)	199
C.8.	Effect of Hiring Rate on Total Schedule Delay (Full Capability Path)	199
C.9.	Effect of Workforce Ceiling on Total Labor Cost (Full Capability Path)	201
C.10.	Effect of Workforce Ceiling on Total Schedule Delay (Full Capability Path)	201
C.11.	Effect of Productivity on Total Labor Cost (Full Capability Path)	202
C.12.	Effect of Productivity on Total Schedule Delay (Full Capability Path)	202
C.13.	Effect of Level of Effort on Total Labor Cost (Limited Capability Path)	205
C.14.	Effect of Level of Effort on Total Schedule Delay (Limited Capability Path)	205
C.15.	Effect of Drumbeat on Total Labor Cost (Limited Capability Path)	206
C.16.	Effect of Drumbeat on Total Schedule Delay (Limited Capability Path)	206
C.17.	Effect of Learning on Total Labor Cost (Limited Capability Path)	207

C.18.	Effect of Learning on Total Schedule Delay (Limited Capability Path)	207
C.19.	Effect of Hiring Rate on Total Labor Cost (Limited Capability Path)	208
C.20.	Effect of Hiring Rate on Total Schedule Delay (Limited Capability Path)	208
C.21.	Effect of Workforce Ceiling on Total Labor Cost (Limited Capability Path)	209
C.22.	Effect of Workforce Ceiling on Total Schedule Delay (Limited Capability Path)	209
C.23.	Effect of Productivity on Total Labor Cost (Limited Capability Path)	210
C.24.	Effect of Productivity on Total Schedule Delay (Limited Capability Path)	210
D.1.	Workforce Profile for One Offshore Patrol Vessel (700,000 Man-Hours, Nine Quarters)	215
D.2.	Aggregate Workforce Profile for Building Offshore Patrol Vessels Starting in 2019, One-Year Drumbeat	217
D.3.	Aggregate Workforce Profile for Building Offshore Patrol Vessels Starting in 2017, One-Year Drumbeat	218
D.4.	Total Labor Costs for Building Ten or 12 Offshore Patrol Vessels, One-Year Drumbeat	219
D.5.	Total Schedule Delay for Building Ten or 12 Offshore Patrol Vessels, One-Year Drumbeat	219
D.6.	Aggregate Workforce Profile for Building Offshore Patrol Vessels Starting in 2019, Two-Year Drumbeat	221
D.7.	Aggregate Workforce Profile for Building Offshore Patrol Vessels Starting in 2017, Two-Year Drumbeat	221
D.8.	Total Labor Costs for Building Ten or 12 Offshore Patrol Vessels, Two-Year Drumbeat	222
D.9.	Total Schedule Delay for Building Ten or 12 Offshore Patrol Vessels, Two-Year Drumbeat	222

Tables

S.1.	Australia's Shipbuilding Industrial Base	xxiv
S.2.	Australia's Ship Repair Industrial Base	xxv
S.3.	Summary Metrics for Australian Shipbuilding Costs Relative to a U.S. Basis	xxxv
2.1.	Australia's Shipbuilding Industrial Base	9
2.2.	Taxonomy of Shipbuilding and Ship Repair Skill Categories	12
2.3.	Australia's Ship Repair Industrial Base	18
2.4.	Current Royal Australian Navy Force Structure	22
3.1.	Projected Work Completion Dates for Current Production Programs	36
3.2.	Two Alternative Royal Australian Navy Acquisition Scenarios	37
3.3.	Alternative Naval Ship Acquisition Strategies Open to Australia	38
4.1.	Peak Workforce Demands for Future Frigate Construction, by Skill Category (Base Case)	55
4.2.	Approximate Skill Percentages of Balanced Workforce Sustainment Levels (Base Case)	56
4.3.	Summary Labor Costs of Various Options for Workforce Sustainment (Full Capability Path)	76
4.4.	Force Structures for Different Drumbeats and Ship Lives (Full Capability Path)	79
4.5.	Future Frigate Force Structure with a Drumbeat of Two, 2026–2041	81
4.6.	Summary Labor Costs of Various Options for Workforce Sustainment (Limited Capability Path)	91

5.1.	Direct Hourly Wage Rates for Boat and Shipbuilding	104
5.2.	Hourly Compensation Costs in Manufacturing (2012)	105
5.3.	Comparison of Physical Characteristics and Hulls Produced, Frigates	110
5.4.	Comparison of Physical Characteristics and Hulls Produced, Destroyers	111
5.5.	Comparison of Physical Characteristics and Hulls Produced, Amphibious Vessels	112
5.6.	Sources of Cost Data, Frigates	113
5.7.	Sources of Cost Data, Destroyers	114
5.8.	Sources of Cost Data, Amphibious Vessels	114
5.9.	Unit Procurement Cost and Relative Index Cost Data, Frigates	117
5.10.	Unit Procurement Cost and Relative Index Cost Data, Destroyers	118
5.11.	Unit Procurement Cost and Relative Index Cost Data, Amphibious Vessels	119
5.12.	Australian Costs to Purchase a DDG-51-Class Destroyer in 2000 and 2014	123
5.13.	Summary Metrics for Australian Shipbuilding Costs Relative to a U.S. Basis	125
5.14.	Number of Months from Keel to Commissioning Table of Means	127
5.15.	Keel-to-Commission Schedule for U.S. Warships	127
A.1.	Australian Naval Ship Production, 1912–1947	155
A.2.	Australian Ship Repair During World War II	157
B.1.	Shipbuilding Workforce Framework	167
B.2.	Summary of Key Variables	169
B.3.	Summary of Baseline Acquisition Scenarios	170
B.4.	Future Frigate Construction Schedule (Base Case: 2020 Start, One-Year Drumbeat)	171
B.5.	Patrol Boat Construction Schedule (Base Case: 2020 Start)	172
B.6.	Littoral Multirole Vessel Construction Schedule (Base Case: 2033 Start)	173
B.7.	Air Warfare Destroyer Construction Schedule (Existing Program)	173
B.8.	Air Warfare Destroyer Construction Schedule (Adding a Fourth Hull)	174

B.9.	Offshore Patrol Vessel Construction Schedule	174
B.10.	Summary of Demand Profile Assumptions	175
B.11.	Productivity as a Function of Experience, by Subcategory	185
B.12.	New Hire Distribution	186
B.13.	Summary of Direct Labor Rates, by Subcategory	186
B.14.	Generic Overhead Rate Assumptions (Full Capability Path)	187
B.15.	Generic Overhead Rate Assumptions (Limited Capability Path)	187
B.16.	Training Costs	188
B.17.	Termination Costs	188
C.1.	Base Case Assumptions for Sensitivity Analysis of Workforce Sustainment Levels	192

Summary

The Australian Department of Defence (AUS DoD) is in the preliminary stages of an ambitious effort to procure up to 50 naval surface warships and submarines over the next two decades. As many as 15 of these vessels might be large surface ships, such as air warfare destroyers (AWDs), landing helicopter docks (LHDs), and Future Frigates, with the remainder being smaller ships, such as patrol boats, offshore patrol vessels (OPVs), and littoral multirole vessels (LMRVs).[1]

This naval demand was first identified in the Australian government's *2009 Defence White Paper* and was refined in the *2013 Defence White Paper*.[2] In 2015, the Australian government will produce a new Defence White Paper to provide a fully integrated and coherent plan for Australia's long-term defense that aligns strategy, capability, and resources. That paper will outline the required structure of the Australian Defence Force and the enablers that are needed to sustain it. The paper will also advise the posture of the Australian Defence Force, in terms of how it works in the region and where in Australia it is located. In preparing this document, policymakers are seeking to gain greater understanding of the ability of Australia's shipyards, workers, and suppliers to produce, deliver, and support naval vessels at the pace and in the order planned by AUS DoD.

---

[1] For the purpose of this analysis, the distinction between patrol boats, OPVs, and LMRVs was used for modeling purposes only. Australia's Force Structure Review process will consider the requirements to address these smaller vessels.

[2] Commonwealth of Australia, *2013 Defence White Paper*, Department of Defence, 2013a, p. 45.

The root question that policymakers are wrestling with is this: Should Australia support a domestic naval shipbuilding industry or buy ships from foreign shipbuilders? The question is complex, and to answer it, policy leaders need to gain an enterprise-level understanding of shipbuilding that brings together the capability requirements, available resources, and the future composition of the Australian shipbuilding and ship repair industrial bases.

To help Australian policymakers gain an up-to-date picture of the country's naval shipbuilding and ship support environment, the AUS DoD's 2015 White Paper Enterprise Management team asked the RAND Corporation in September 2014 to conduct a three-pronged analysis that would:

- provide an understanding of Australia's current shipbuilding capabilities and gauge how alternative acquisition strategies might affect both the capacity of the domestic industrial base and the total cost of the enterprise
- compare the costs of Australia's naval shipbuilding industry with overseas manufacturers that produce platforms of comparable size and scope
- assess the economic costs and benefits of government investments in Australia's naval shipbuilding industrial base under the various enterprise options.

Relying both on public and proprietary data and on surveys of industry representatives, RAND analyzed these issues between September 2014 and March 2015.

Shipbuilding and Ship Support Industrial Bases and Programs

Historically, Australia has acquired warships from overseas—for example, the *Charles F. Adams* guided missile destroyer and the first four *Oliver Hazard Perry* guided missile frigates. More recently, the focus has shifted to acquiring ship designs from overseas, modifying them to meet Australian requirements, and using them to build all or parts of warships in Australia. Once they have entered service, Australia's warships traditionally have been maintained and serviced in-country.

This pattern points to two parallel and overlapping industrial bases in Australia: one for shipbuilding and another for ship maintenance, repair, and sustainment. With respect to naval shipbuilding, four main companies comprise Australia's industrial base today: ASC Pty Ltd, with headquarters in Adelaide, South Australia; Austal, with headquarters in Perth, Western Australia; BAE Systems Australia, a subsidiary of BAE Systems plc, with headquarters in Adelaide, South Australia; and Forgacs, with headquarters in Newcastle, New South Wales. Those four companies—along with Defence Maritime Systems, Naval Ship Management (Australia) Pty. Ltd., and Thales Australia—constitute the bulk of Australia's naval ship repair and support industrial base. Table S.1 and Table S.2 list the companies' facilities and current programs related to shipbuilding and ship repair, respectively.

The tables reveal that maintenance and production demands are distributed across a relatively large number of organizations: seven (not including subcontractors) constitute the Australian shipbuilding and ship repair industrial base today. Moreover, while Australia's maintenance and production activities are distributed across similar companies, they take place at different shipyards. All the organizations that have a role in production also have maintenance contracts. However, in general, shipyards that support shipbuilding do not support maintenance and repair activity.

Australia has a workforce of several thousand workers who have experience directly relevant to shipbuilding and ship support. RAND's survey of the industry revealed that a total of 7,950 employees were working across shipbuilding and submarine and ship repair in 2013–2014, with ASC accounting for approximately one-half of that total.[3] About 4,000 of the industry total were associated with current shipbuilding projects, with the rest being in ship support. While these war-

---

[3] These employment figures differ from those shown in Commonwealth of Australia, *Future Submarine Industry Skills Plan*, Department of Defence, Defence Materiel Organisation, 2013b. This difference is mainly a result of that report including only workers involved with producing the AWDs, LHDs, and patrol boats. RAND's figures include all workers involved in production, plus those workers involved in maintenance and modernization.

Table S.1
Australia's Shipbuilding Industrial Base

Company	Shipyards Currently in Use	Current AUS DoD Shipbuilding Programs	Other Current Shipbuilding Programs
ASC	ASC South Adelaide, South Australia[a] South Australian government's Common User Facility (CUF)[b]	Hobart-class AWD (block construction and integration)	None
Austal	Austal Henderson, Western Australia[c]	None	Customs and Border Protection Services patrol boats
BAE Systems Australia	BAE Williamstown, Victoria[d]	Hobart-class AWD (block construction) Canberra-class LHD (system installation and final outfitting)	None
Forgacs	Tomago, New South Wales[e]	Hobart-class AWD (block construction)	None

[a] Located northwest of Adelaide, ASC South is used for the consolidation of the AWDs and is Australia's largest naval shipbuilding hub, incorporating a critical mass of warship design and construction skills. ASC North is a high-tech submarine maintenance facility containing shiplift, docking, and transfer system halls; wharf facilities; warehousing; hardstand area; a dedicated painting and blasting facility; and construction and assembly halls. See ACIL Allen Consulting, *Naval Shipbuilding & Through Life Support, Economic Value to Australia*, ACIL Allen report to Australian Industry Group, December 2013.

[b] The CUF is a national strategic asset owned and operated by the government of South Australia. The facility is spread across eight hectares at the heart of Techport, Australia, northwest of downtown Adelaide. See ACIL Allen Construction, 2013.

[c] Located at Cockburn Sound, south of Perth, the facility integrates naval shipbuilding and ship repair with facilities that also support the oil and gas sectors at the Australian Maritime Complex CUF in Western Australia. This CUF was jointly funded by the federal and Western Australia governments to assist local industry with competing for services to the oil and gas, resources, and marine/defense industries. See ACIL Allen Construction, 2013.

[d] Williamstown is situated on the western shore of Port Phillip Bay. It has two building berths; travelling cranes; a graving dock; transporters with large capacity; halls for module construction, assembly, and blast, paint, and outfit; and fully serviced wharfage. Williamstown received a substantial capital upgrade to assist with the AWD project. See ACIL Allen Construction, 2013.

[e] This nine-hectare waterfront shipyard on the Port of Newcastle hosts Forgacs' build of AWD modules. It also is used for marine vessel conversions, refits, unscheduled repairs and maintenance, and survey dockings, with two slipways and extensive wharfage. See ACIL Allen Construction, 2013.

Table S.2
Australia's Ship Repair Industrial Base

Company	Shipyards Currently in Use	Current AUS DoD Ship Repair Programs
ASC	• ASC North, Adelaide, South Australia • Australian Marine Complex CUF, Henderson, Western Australia • Fleet Base West, Garden Island, Western Australia • Fleet Base East, Garden Island, New South Wales	• *Collins*-class submarine
Austal	• Henderson, Western Australia • Darwin Naval Base	• *Armidale*-class patrol boats subcontractor to Defense Maritime Systems
BAE Systems Australia	• Henderson, Western Australia	• *Anzac*-class frigate modernizations
	• HMAS *Cairns*, Cairns, Queensland	• Hydrographic ships
	• HMAS *Waterhen*, New South Wales	• Minehunter coastal
		• *Canberra*-class LHD
	• Williamstown, Victoria	• *Anzac*-class frigate modernizations
Defence Maritime Systems[a]	• HMAS *Coonawarra*, Darwin, Northern Territory • HMAS *Cairns*, Cairns, Queensland	• *Armidale*-class • Patrol boats
Forgacs	• Carrington, New South Wales • Fleet Base East, Garden Island • New South Wales	• Replenishment ships • *Tobruk*-class landing ship heavy
NSM	• Henderson, Western Australia • Fleet Base East, Garden Island, New South Wales • HMAS *Cairns*, Cairns, Queensland	• *Anzac*-class frigate maintenance and support
Thales	• Fleet Base East, Garden Island, New South Wales	• *Anzac*-class frigate • Combat system support • Guided missile frigate • support and upgrades • *Tobruk*-class landing ship heavy

[a] Defense Maritime Systems manages repair of survey ships at HMAS *Coonawarra*, Darwin, Northern Territory, and HMAS *Cairns*, Cairns, Queensland.

ship building and support workforce headcounts do not include subcontractors, Australia's overall numbers by category appear small in comparison to other large shipbuilding projects conducted overseas.

As Table S.1 shows, Australia has two active Royal Australian Navy (RAN) shipbuilding projects: the LHD and the AWD. The *Canberra*-class LHD project is a two-ship class whose hulls have been built by the Spanish firm Navantia and shipped to Australia, whereupon their superstructures have been fabricated, equipped, and outfitted by BAE Systems Australia.[4] At about 230 m in length and with a displacement of 27,500 metric tons, the LHD is the largest vessel in RAN. Construction of the first of class, the HMAS *Canberra*, began in 2008, and the hull was launched in 2011. Work on the second LHD, the HMAS *Adelaide,* began in 2010; it was launched in 2012.[5]

The *Hobart*-class AWD is slated to replace RAN's *Adelaide*-class frigates. Current plans call for a three-ship class.[6] Their production is being overseen by a consortium made up of the Defence Materiel Organisation, ASC, and Raytheon, known as the AWD Alliance. The *Hobart*'s keel was laid in 2012 and the *Brisbane*'s in 2014. Originally,

---

[4] Outfitting tasks occur either during the construction of the pieces that make up the ship or when those pieces are assembled to form the completed ship. Outfitting covers a broad range of functional tasks including:

- structural: installing equipment foundations, doors, ladders, hatches, and windows
- piping: installing and welding pipes, including spools and connectors
- electrical power distribution: installing the power distribution system downstream of the main power switchboards, including hanging and pulling cables and installing local switchboards and ancillary electrical equipment
- Heating, ventilation, and cooling (HVAC): installing air handling units, ducting, and other ancillary HVAC equipment
- joinery: installing accommodations, such as cabins or berths, dining facilities, food preparation areas, and rooms for meetings or other administrative purposes
- painting and insulation: covering the structure and accommodations of the ship.

For naval combatants, outfitting also includes the installation of combat and weapon systems.

[5] This information was drawn from Naval Technology, "Canberra Class Landing Helicopter Docks (LHDs), Australia," web page, undated(a); IHS, *Jane's Fighting Ships* (online), undated(a); and other publicly available sources.

[6] There had been a contractual option for a fourth AWD, but that has expired. Our analysis in subsequent chapters assumes that if that fourth AWD were built, it would be the same design as the previous three AWDs.

the new destroyers were to be operational between 2014 and 2017, but those dates have slipped to 2016 and 2019.[7]

Emerging Short-Term and Longer-Term Demand Gaps

ASC, BAE Australia, and Forgacs are rapidly approaching the end of their work on the AWD and LHD.[8] Forgacs will finish its AWD blocks around the third quarter of 2015. BAE Australia finishes work on the LHD and AWD around the third quarter of 2015 and the second quarter of 2016, respectively, although AWD work will start to decrease significantly about a year before that. ASC's structural and outfitting work on the AWD will begin to decrease in 2017 and will be completely finished in 2019. The three shipbuilders have already started to shed workforce and, barring new programs, will have no more structural or outfitting work after the second quarter of 2019.

Other naval shipbuilding efforts are on the horizon. Australia intends to build Future Frigates, which will replace soon-to-retire *Anzac*-class frigates and provide greater antisubmarine capabilities. Additionally, it has plans to acquire a fleet of OPVs. And the *2013 Defence White Paper* stated that Australia will need to replace its *Collins*-class submarine fleet with 12 new-generation submarines.[9]

However, the timing of these planned acquisitions is likely to produce short-term and long-term gaps in demand for shipyard production facilities, services, and workers. Our analysis suggests that in the short term, a gap will appear between the end of the AWD production and the start of the Future Frigate program. Another gap will arise around 2035, when production of the Future Frigate is expected to end.

Figure S.1 shows our projection of the gaps' timing and size under Australia's current acquisition plan, measured in terms of the shipbuilding industry's full-time-equivalent (FTE) headcount over the next two

---

[7] This description was drawn from IHS (undated) and other publicly available sources.

[8] Austal is currently building patrol boats for the Australian Customs and Border Protection Service, but that work is also nearing completion.

[9] Commonwealth of Australia, 2013a, pp. 79–83.

Figure S.1
Workforce Profile for Building Air Warfare Destroyers and Future Frigates (Base Case)

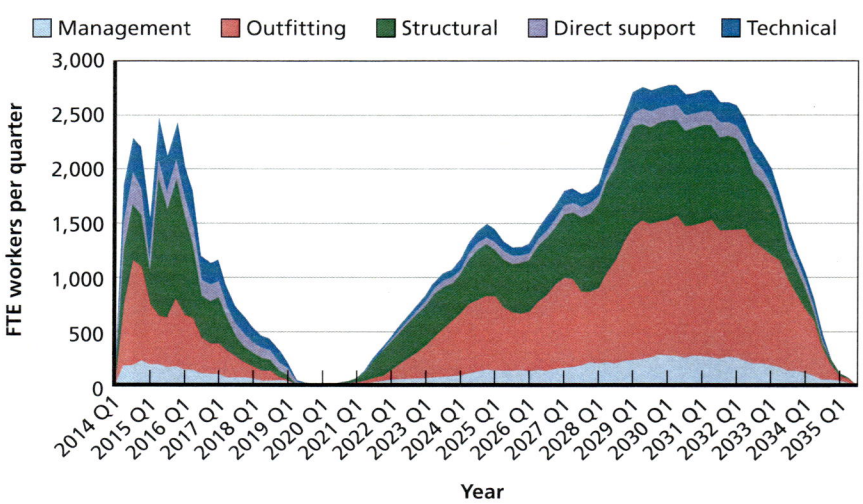

RAND RR1093-S.1

decades. AWD work, which is reflected in the profile on the left side of the graph, will end in the next one to three years; as it declines, shipyards—especially those building blocks for the AWD—will begin to shed their workforces. Because construction of the Future Frigate will not start until 2020, reflected in the right side of the graph, there is the potential that demand for workers could fall to zero, with reverberations that may last three to five years after Future Frigate production ramps up.[10] Without some way to lessen the gap between the end of the AWD program and the start of building the Future Frigate, the industrial base will have to ramp up its workforce from an almost negligible level to 2,700 skilled personnel in approximately eight years.

---

[10] We assume that the first-of-class Future Frigate will take 6.5 years to build and will require 5.5 million fully productive man-hours. The second ship in the class will start in 2023, the third ship will start in 2025, and the remaining ships will start one per year after that (approximately matching 30 years from the *Anzac*-class in-service dates). We assume that the second ship will require 5 million fully productive man-hours, and subsequent follow-on ships follow a 95-percent unit learning curve.

Options for Lessening the Short-Term and Long-Term Gaps

Using an adaptable shipbuilding and force structure modeling tool that RAND has employed on numerous industrial base studies, the project team analyzed a variety of strategies that the Australian government could employ to close the short-term gap in demand for naval shipyard workers while retaining the ability to build all the planned warships domestically, including the following:

- Start construction of the Future Frigate class before 2020. Although it would be unlikely, if a Future Frigate design could be ready and construction started by 2018, the shipyard employment would rise from zero to roughly 200 workers during the gap period, or a bit less than 10 percent of the peak of the Future Frigate program.
- Build a fourth AWD. With a timely award of the fourth ship, shipbuilders could transition from finishing the third ship to beginning the fourth and, with proper planning, could sustain their workforces of up to 400–900 shipyard employees until the Future Frigate program begins.[11] However, there is no stated requirement for a fourth AWD.
- Build patrol boats in the major shipyards in addition to Future Frigates. To lessen the gap, the shipyards that build large combatants could start building patrol boats in 2017, in the period between the end of the AWD program and the start of the Future Frigate build (which could be in either 2018 or 2020). This could keep up to 200 employees working at shipyards during the gap.
- Build OPVs in the major shipyards in addition to Future Frigates. If Australia were to start construction of OPVs by the end of 2017, between 400 and 500 shipyard workers could be retained throughout the gap years separating the end of AWD construction and the commencement of the Future Frigate program.

---

[11] We assumed that building the fourth AWD reduces the Future Frigate buy to seven ships and that it would replace the first *Anzac* ship planned to retire in 2026.

Figures S.2 and S.3 display two aspects of the strategies we examined (juxtaposed next to Australia's current baseline plan, portrayed in the red bars, of producing eight Future Frigates beginning in 2020). Figure S.2 compares the costs of the strategies. It shows that all would cost some AUD 5.5 billion, strongly suggesting that keeping as much as 30 percent of shipyards' Future Frigate workforces employed during the gap years would not be much costlier than allowing worker headcounts to drop to zero. In addition, it shows that lessening the gap by building OPVs (portrayed in the cross-hatched bar) would provide additional ships to RAN at a very marginal labor cost to produce them. Figure S.3 shows that most options for lessening the gap would significantly reduce the total delay in delivering *Anzac*-class replacements.

Australia could turn to several continuous build strategies to sustain a cost-effective domestic shipbuilding industrial base. Start-

Figure S.2
Total Labor Costs of Base Case and Alternative Shipbuilding Construction Paths

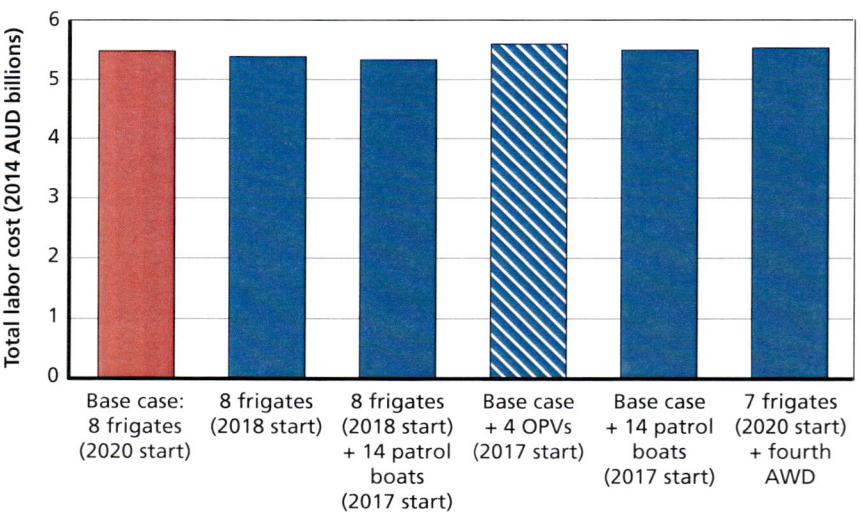

NOTE: The figure assumes that the base case Future Frigate uses 5 million man-hours, takes 6.5 years to build, and has a 95-percent unit learning curve; in addition, the last six ships are produced with a drumbeat of one.
RAND RR1093-S.2

Figure S.3
Total Schedule Delay of Base Case and Alternative Shipbuilding Construction Paths

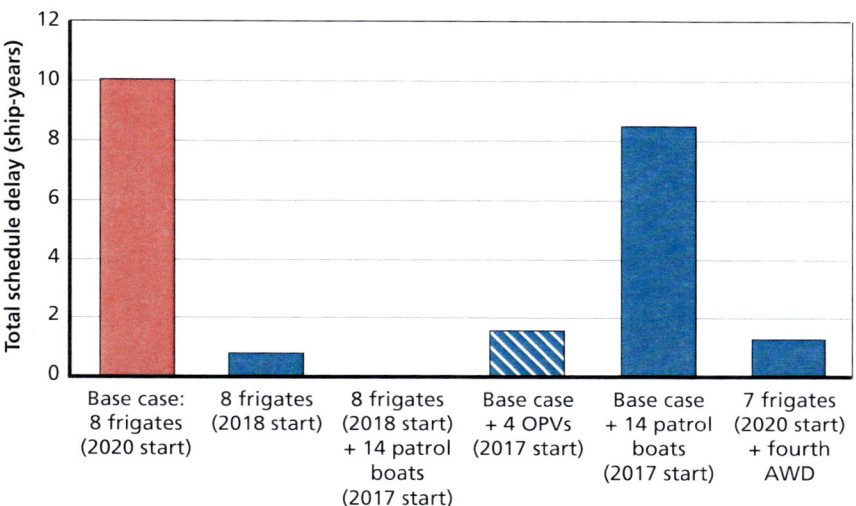

NOTE: The figure assumes that the base case Future Frigate uses 5 million man-hours, takes 6.5 years to build, and has a 95-percent unit learning curve; in addition, the last six ships are produced with a drumbeat of one.
RAND RR1093-S.3

ing construction of Future Frigates every one-and-a-half to two years beginning with the third vessel in the class would sustain a skilled workforce beyond the Future Frigate program.[12] However, starting construction of the last six Future Frigates at that pace will result in the delay of new ships to replace the *Anzac*-class ships as they retire. This delay may require that the operational lives of the *Anzac* ships be extended beyond 30 years or present a shortfall in RAN major warships for several years. Alternatively, Australia could maintain a drumbeat of one for the Future Frigate and begin producing smaller LMRVs as the Future Frigate build ends.

---

[12] The pace of ship construction is also called a *drumbeat*. From a shipbuilding perspective, the drumbeat refers to how frequently new ships are delivered to RAN. For example, a drumbeat of one implies that a new ship is delivered each year; a drumbeat of two implies that a new ship is delivered at two-year intervals.

It may be difficult for Australia to sustain more than one domestic shipbuilder of large warships in the short term. However, there is the risk that a natural or man-made disaster could shut down that one shipyard for a period of time. As noted, the various future acquisition options shown in Figures S.2 and S.3 would generate a total demand for approximately 2,700 workers, maybe up to 4,000 under certain Future Frigate man-hours-per-ship workload assumptions. This demand could support a shipyard that builds blocks and a shipyard that builds and assembles blocks. There are costs and risks of having more than one shipbuilder (such as inefficiencies in labor, excess costs in overhead, and scheduling problems), but if the national decision is to have two shipbuilders, adequate productive work must be assigned in the workforce demand gap. The anticipated future workforce demands make it difficult and costly to sustain more than two shipyards.

We also examined the implications for the short-term gap in workforce demand if Australia were to move toward an industrial base that built some, not all, of the planned RAN warship acquisitions or that only outfitted ships that are built overseas. The problems and challenges that the industrial base would encounter are broadly similar to ones faced by a fully indigenous industry. However, the gap will be two to three years longer, because the first of the Future Frigates would begin construction overseas, and outfitting skills will be more important to sustain than structural skills. The candidate solutions and options are also similar: Starting the Future Frigate earlier, in 2018, would reduce costs and delays; adding patrol boats or OPV could help, if they were built at the same shipyard as the Future Frigate; and adding a fourth AWD could reduce delays (although it would increase costs). However, in comparison with a fully capable industrial base, more work will be needed to sustain the workforce in the longer gap period, and lessening the gap by producing full ships (e.g., a fourth AWD, patrol boats, or OPVs) would sustain structural skills that are unneeded to outfit the Future Frigates. Moreover, future demands on an industrial base that builds only portions of a ship would make it difficult to support more than one shipbuilder.

In the long term, a continuous build strategy of building major surface combatants with a drumbeat of 1.5 to two should sustain a healthy and cost-effective shipbuilding industrial base. Building OPVs during the short-term gap will provide a cost-effective transition to the lower demands of a Future Frigate program resulting from a drumbeat greater than one. But adopting this option will present challenges with a number of pre-conditions required to achieve it, including starting production by 2017, using an existing design without modifications, and strategically scheduling the build program to complement the Future Frigate workforce profile. And the end of the Future Frigate build program would flow into the build of the next major surface combatant. There will be challenges during the replacement of the *Anzac* class, but these challenges might be met with careful planning of delivery schedules and extended usage of the existing fleet.

How Does Australia's Shipbuilding Industry Compare with Comparable Overseas Producers?

RAND compared the relative performance of Australian industry with other naval shipbuilding nations. At the heart of this inquiry is the question of whether Australia pays a premium for its indigenously built naval vessels, and if so, how large that premium is.

To pursue this line of inquiry, we focused on benchmarking the industry's performance. Benchmarking is the process of comparing the performance and practices of one firm, country, or system with another, at either an aggregate or unit level (e.g., program or item). This comparative process is used frequently in the commercial sector to identify strengths, weaknesses, and areas for improvement. Benchmarking is often focused on identifying best practices and their degree of implementation across the comparison organizations.

We focused on comparing Australia's shipbuilding industry with its overseas peers on two dimensions: cost and schedule.

Comparing Australian Industry Costs with Overseas Peer Industries

We relied on the following three benchmarks to analyze cost data:

1. Input benchmarking, which uses inputs to shipbuilding, such as labor costs and material costs, to project relative shipbuilding costs between countries

2. Comparative benchmarking, which compares similar systems directly in terms of cost performance on a cost-per-metric-ton (CPT) basis
3. Parametric benchmarking, which is a statistical method that defines a baseline (or typical) performance based on key system characteristics (e.g., displacement weight and speed).

After attaching metrics to each of these benchmarks, we found that Australian naval shipbuilding tends to be more expensive than a variety of comparator countries: Italy, Japan, Republic of South Korea, Spain, the United Kingdom, and the United States. The ships that we used in the comparative benchmarking aspect of this analysis are shown in Figure S.4. Table S.3 displays the metrics that we associated

Figure S.4
Ships Used in Comparison Benchmarking

Frigates
F590 FREMM (France)
D650 FREMM (Italy)
LCF (Netherlands)
Iver Huitfeldt (Denmark)
Anzac (Australia)
Incheon (Republic of South Korea)
FFG-7 (United States)
LCS (United States)

Destroyers
JDS Akizuki (Japan)
JDS Atago (Japan)
Hobart (Australia)
DDG-51 (United States)
Sejong-Daewang KDX-3 (Republic of South Korea)
Type 45 (United Kingdom)
F105 (Spain)

Amphibious
Canberra (Australia)
Juan Carlos (Spain)
LHD-1 (United States)
LPD-R (United Kingdom)
JDS Izumo (Japan)

SOURCE: Royal Australian Navy photos.
NOTE: FREMM = frigate or frégate multimission; LCF = air defense and command frigate; FFG = guided missile frigate; LCS = littoral combat ship; JDS = Japan Defense Ship; DDG = guided missile destroyer; KDX = Korean Destroyer Experimental; LHD = landing helicopter dock; LPD = landing platform/dock.
RAND RR1093-S.4

Table S.3
Summary Metrics for Australian Shipbuilding Costs Relative to a U.S. Basis

Method	Metric	Approximate Australian Premium Relative to a U.S. Basis (%)
Input	Direct shipbuilding labor wages	40
	Manufacturing labor costs	35
	Oil and gas industry construction	20
	Construction cost adjusted to First Marine International shipbuilding productivity[a]	45
Comparative	Frigate costs	40
	Destroyer costs	30[b]
	Amphibious ship costs	12[c]
Parametric		35

[a] Cost comparison based on hours per compensated gross tonnage, a productivity measure used by First Marine International Ltd. This measure compares the weights of different types of commercial ships with one another by using adjustment factors that depend on the ship type (e.g., tanker, dry cargo, ferry). See First Marine International Ltd., *First Marine International Findings for the Global Shipbuilding Industrial Base Benchmarking Study*, Part 2: *Mid-Tier Shipyards, Final Redacted Report*, February 6, 2007.

[b] Prior to rebaseline.

[c] Based on the recent LHD. Because significant portions of the ship are built in Spain, the relative costs may not be representative of a complete Australian build (the premium is likely lower than if the ship had been fully built in Australia).

with the three benchmarks and summarizes Australia's premium relative to a U.S. basis for those metrics.

Table S.3 shows that relative to U.S. shipbuilding costs, the premium for ships entirely built in Australia ranges from 30 percent to 45 percent. For ships built partially in Australia, this premium is lower. Combatants (frigates and destroyers) seem to have a consistent premium of around 30 percent to 40 percent. The premium for amphibious ships is lower, but it is still some 12 percent more than a U.S. basis.[13]

---

[13] Note that the CPT metric is less robust for amphibious ships and reflects that a significant portion of the ship has been built in Spain.

Overall, the three benchmarking methods indicate a modal premium of about 30 to 40 percent for naval warships built entirely in Australia. This perceived premium, it should be noted, can be significantly influenced by foreign exchange rates, and any consideration of foreign or domestic build must take into consideration currency exchange factors and risks.

Comparing Australian Industry Schedules with Overseas Peer Industries

To explore the schedules of Australian-built warships with those built by overseas producers, we examined the number of months it has taken the European, Japanese, Korean, and U.S. producers of the warships used in our comparison to produce the vessels, as measured from laying the keels for the warships to commissioning them. We then compared those schedules with the length of time it has taken two Australian ship classes, the *Anzac* and the AWD, to reach those milestones. See Figure S.5.

Figure S.5
Average Keel to Commission Schedule

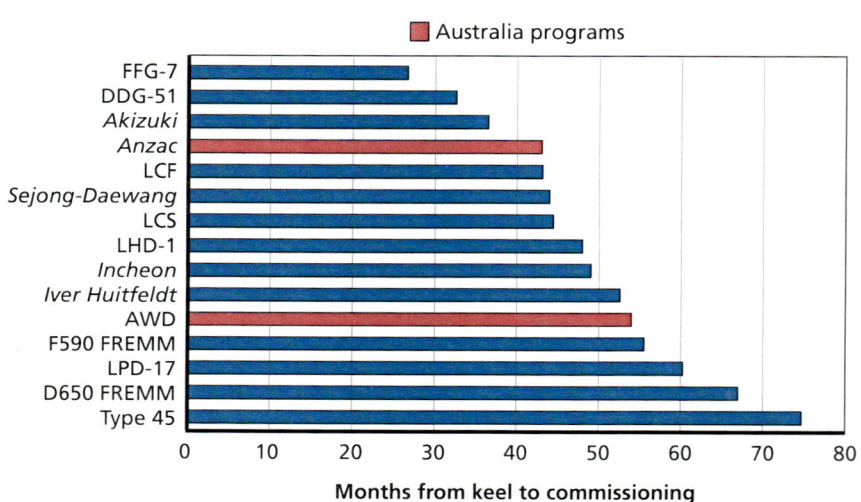

Overall, we found that the average time of the Australian *Anzac* class to go from laying the keel to commissioning has been slightly faster than the average of the comparator ships produced overseas; the average time for the AWD has been on par with the average of the overseas comparators.

Can the cost and schedule performance of Australia's shipbuilding industry improve? Our benchmarking analysis, particularly our examination of input benchmark factors, suggests that the domestic industry's cost premium can be reduced, but only if Australia is able to implement two changes. It will need to institute and adhere to a consistent production pace and demand for warship builds; in other words, it will need to adopt steady and predictable production drumbeats for future acquisitions. At the same time, Australia will need to embrace and institutionalize the modified acquisition practices outlined in this report, such as what was seen in the United States' submarine-building industry at the end of the Cold War. The industry recognized that its products were becoming unaffordable and made radical changes to the way it designed and built submarines, focusing on cost-effectiveness. Very strong and visionary leadership at the companies drove this change. Australia will also need to have mature designs in hand when construction on a warship begins, make minimal changes once production begins, and maintain a fully integrated team—comprising the designer, builder, and suppliers—from start to finish. Such changes will not happen overnight, however.

What Are the Economic Costs and Benefits of Government Investments in Australia's Naval Shipbuilding Industrial Base?

In this component of the analysis, the RAND team assessed the relationship between AUS DoD's maritime spending and levels of output, employment, and earnings. The team looked for possible favorable effects, such as the economic multiplier effect that such spending would have on Australia's support and supply chain, and spillover effects, such as new technologies or skill developments. At the same time, it looked for possible unfavorable effects, such as defense spending crowding out other economic development.

The team conducted an extensive review of the academic and business literature on economic multipliers. It also examined three cases of military spending—the effect of shipbuilder Austal USA on the economy of Mobile, Alabama; the effect of Newport News Shipbuilding on the economy of Newport News, Virginia; and the effect of the Saab Aeronautics program to construct the Gripen jet fighter on Sweden's economy.

The literature search uncovered no consensus on the effect of military spending on local and regional economies. However, the case studies suggested that spending on naval shipbuilding can have favorable local and regional effects, especially during times of overall economic distress. But those effects are localized to a large degree, and it is unrealistic to expect that shipbuilders will produce significant favorable spin-offs and spillovers. Sweden's experience with the Gripen, which spawned ancillary jobs and start-up companies, was not seen to have been replicated in the U.S. military shipbuilding environment and, as such, might be an overly optimistic analogy for Australia's industry.

Wrapping Up: How Could Australia Sustain a Naval Shipbuilding Industry?

Four overarching findings emerge from this analysis of Australia's naval shipbuilding industrial base:

1. Production of naval warships in Australia involves a 30 percent to 40 percent price premium over the cost of comparable production at shipyards overseas. This premium could drop over time, however, with steady production drumbeats and mature designs.

2. The economic benefits of a domestic naval shipbuilding industry are unclear and depend on broader economic conditions. That said, the industry could potentially employ more than 2,000 people in long-term positions.
3. Controlling critical production offers wider strategic benefits and flexibility. It would avoid dependence on foreign sources; enable performance of ship alterations, modernizations, and life-of-class maintenance; and support in-country suppliers.
4. Sustaining a naval shipbuilding industry will require specific steps. These include adopting a continuous build strategy starting with the Future Frigate and matching industrial base structure to demand.

Acknowledgments

This work could not have been undertaken without the special commitment and support that the Australian Department of Defence White Paper Enterprise Management team provided to RAND. For that support over more than six months of intensive study and analysis, we are grateful. Many individuals throughout the Australian Department of Defence and in Australian industry provided data and shared insights that were critical to our quantitative analysis and to the interpretations and conclusions described in this report. Their names and contributions would fill several pages.

We particularly wish to recognize and thank five individuals without whose broad-based participation and support this analysis would not have been possible: Marc Ablong, first assistant secretary, White Paper; Kate Louis, assistant secretary, White Paper Enterprise Management; Darren Sutton, specialist advisor, White Paper Enterprise Management; Glenn Alcock, assistant director, White Paper Enterprise Management; and Cameron Gill, director, White Paper Enterprise Management. We appreciate their efforts on our behalf and the friendships that ensued.

We wish to thank the Australian firms and organizations that hosted our visits and provided data and insights: Anzac Ship Alliance, ASC, Austal, BAE Systems, DMS Marine, Forgacs, Naval Ship Management, and Thales.

We would also like to thank Professor Thomas Lamb, University of Michigan, College of Engineering, Department of Naval Architecture and Marine Engineering, for the insights and data he shared with us.

RAND colleague Jessie Riposo provided a thoughtful review of a draft of this report, as did RDML Joe Carnevale (Ret.). Their careful reads of our document and insightful comments occasioned many changes that improved the substance and clarity of the final product.

Lastly, the authors owe RAND colleagues Jack Riley, Cynthia Cook, Nancy Pollock, Michelle Platt, and Joan Meyers an incalculable debt for their thorough and patient assistance at every stage in the project. Allison Kerns elegantly edited the document under considerable time constraints.

Abbreviations

ADI	Australian Defence Industries
AMEC	Australian Marine Engineering Corporation
AUS DoD	Australian Department of Defence
AUD	Australian dollar
AWD	air warfare destroyer
CAD	computer-aided design
CGT	compensated gross tonnage
CPT	cost per metric ton
CUF	Common User Facility
DDG	guided missile destroyer
DMO	Defence Materiel Organisation
FF	Future Frigate
FFG	guided missile frigate
FREMM	frigate or frégate multimission
FTE	full-time equivalent
FY	fiscal year
HM&E	hull, mechanical, and electrical

HMAS	Her or His Majesty's Australian Ship
HVAC	heating, ventilation, and cooling
IH	*Iver Huitfeldt*
JDS	Japan Defense Ship
KDX	Korean Destroyer Experimental
LCC	life-cycle cost
LCF	air defense and command frigate
LCS	littoral combat ship
LHA	landing helicopter assault
LHD	landing helicopter dock
LMRV	littoral multirole vessel
LPD	landing platform/dock
LSD	landing ship dock
LSH	landing ship heavy
MOTS	military off-the-shelf
NNS	Newport News Shipbuilding
NSM	Naval Ship Management (Australia) Pty. Ltd.
OMT	Odense Maritime Technology
OPV	offshore patrol vessel
PB	patrol boat
RAN	Royal Australian Navy
USD	U.S. dollar
WBS	work breakdown structure

CHAPTER ONE

Introduction

The Australian Department of Defence (AUS DoD) is in the early stages of an ambitious effort to procure up to 50 naval surface ships and submarines over the next two decades. This includes between 13 and 15 large surface ships, such as air warfare destroyers (AWDs), landing helicopter docks (LHDs), and Future Frigates, and between 27 and 35 smaller ships, such as patrol boats, offshore patrol vessels, and littoral multirole vessels. In the process, defense policymakers are seeking to gain greater understanding of the ability of Australian shipyards, workers, and suppliers to produce, deliver, and sustain those vessels at the pace and in the order planned by AUS DoD. To inform this process, it was necessary to conduct our analysis in parallel with the ongoing development of the Force Structure Review. The resulting demand profiles used in this report were therefore used as exemplars as the government considers its final force structure requirements through the White Paper process.[1]

This naval demand was first identified in the Australian government's *2009 Defence White Paper* and was refined in the *2013 Defence White Paper*.[2] In 2015, the Australian government will produce a new Defence White Paper to provide a fully integrated and coherent plan for Australia's long-term defense that aligns strategy, capability, and

---

[1] For the purpose of this analysis, the distinction between patrol boats, offshore patrol vessels, and littoral multirole vessels was used for modeling purposes only. Australia's Force Structure Review process will consider the requirements to address these smaller vessels.

[2] Commonwealth of Australia, *2013 Defence White Paper*, Department of Defence, 2013a, p. 45.

resources. That paper will outline the required structure of the Australian Defence Force and the enablers that are needed to sustain it. The paper will also advise the posture of the Australian Defence Force, in terms of how it works in the region and where in Australia it is located. In preparing this document, policymakers are seeking to gain greater understanding of the ability of Australia's shipyards, workers, and suppliers to produce, deliver, and support naval vessels at the pace and in the order planned by AUS DoD. This work will inform an enterprise-level shipbuilding plan that brings together navy capability requirements, available resources, and the future composition of the Australian shipbuilding and ship repair industrial bases.

As it prepares the new Defence White Paper, a basic question facing the government is whether Australia—to fulfill the acquisition program that defense leaders laid out in 2013 and that may be expanded upon in 2015—should support a domestic naval shipbuilding industry or buy ships from foreign shipbuilders. This question is complex, containing many facets and issues that often center on cost trade-offs and economic considerations but that also touch upon important national and strategic concerns. If policymakers desire to maintain an indigenous Australian shipbuilding industry, they need to address a range of follow-on questions about what future demands must be met to permit the industrial base to operate effectively and efficiently and how the assets of the shipbuilding and ship repair industrial bases should be organized.

RAND Research Objective

At the request of AUS DoD's 2015 White Paper Enterprise Management team, the RAND Corporation has been analyzing the capability of the shipbuilding and ship repair industrial bases in Australia to meet the demands of current and future naval surface ship programs. RAND's analysis, conducted between September 2014 and March 2015, focuses on answering the following fundamental questions related to the ability of Australia's naval shipbuilding industrial base to successfully implement the current acquisition plan for surface warship production and sustainment:[3]

- What are the comparative costs associated with alternative shipbuilding paths?
- Is it possible for Australia's naval shipbuilding industrial base to achieve a continuous build strategy, and how would the costs of such a strategy compare with the current and alternative shipbuilding paths?
- How do the costs of acquiring vessels domestically compare with the costs of acquiring comparator(s) from shipbuilders overseas?
- How much do expenditures connected with warship building, maintenance, and sustainment add to Australia's economy?

Relying both on public and proprietary data and on surveys of industry representatives, the analysis addresses these questions by examining the capacity of the current workforce and facilities of the Australian industrial base, identifying demands for those resources during the next two decades, and exploring options to address situations in which future demands might exceed available capabilities. The study aims to help Australia's defense policymakers in three ways: first, to gain an understanding of the capacity and associated costs of Australia's naval shipbuilding industrial base to successfully implement AUS DoD's current acquisition plan; second, to gauge how alternative acquisition requirements, programs, build strategies, quantities, and related costs and schedules might affect the capacity of that industrial base; and third, to measure the economic impact of the industry throughout Australia.

---

[3] Readers should note that this analysis focuses solely on the industrial base responsible for producing and sustaining surface vessels. While submarine production and sustainment relies on some of that industrial base, our charter from AUS DoD was to restrict our examination to the industries upon which naval surface forces depend.

Structure of the Report

RAND answered the four research questions listed above in the context of alternative acquisition strategies. We did so by seeking to understand the capacity of Australia's naval shipbuilding and ship repair industrial bases to successfully implement the current acquisition plan for surface warship production and sustainment, by comparing the costs of Australia's naval shipbuilding industry with overseas manufacturers that produce platforms of comparable size and scope, and by assessing the benefits and costs of government investments in Australia's naval shipbuilding industrial base under the various enterprise options.

Following this introduction, Chapter Two describes how naval shipbuilding and ship support are different from other major industries and how they differ from each other. We also provide a snapshot of the organizations that are part of the Australian naval ship industrial bases. We then provide the background for our assessment that the in-service ship support industrial base is fairly robust and will not face the types of challenges and major decisions that will arise in naval shipbuilding. In Chapters Three through Six, we concentrate on the shipbuilding industrial base. In Chapter Three, we provide details on the current and planned Australian naval ship acquisition plans and describe how current plans lead to short-term and long-term demands for shipbuilding resources. We define several future acquisition paths that Australian naval shipbuilding can take and describe general options along those paths. Chapter Four provides our analytical assessment of the direct shipbuilding labor costs of adopting different strategies and acquisition options for the future Australian shipbuilding enterprise. Chapter Five provides a comparison of Australian naval shipbuilding costs with those of other nations. These comparisons help to determine general cost trade-offs of foreign and domestic shipbuilding. Chapter Six describes the potential economic spin-off of spending shipbuilding dollars in Australia versus overseas. Chapter Seven synthesizes the findings from the overall research and provides an initial set of findings and recommendations. Several appendixes provide more details on various aspects of the research.

CHAPTER TWO

Australia's Naval Shipbuilding and Ship Repair Industrial Bases

This chapter discusses the unique nature of naval shipbuilding and ship repair industries and provides a brief overview of those operating in Australia. We provide a snapshot of the ability of the industrial base to accommodate Australia's naval shipbuilding program in the near and longer terms. To some extent, this ability is colored and constrained by the history of naval shipbuilding over the past few decades. That history has been marked by an inconsistent and unstable business environment for naval surface ship production for more than 50 years.

Historically, Australia has acquired military off-the-shelf (MOTS) ship designs from other countries, modifying them to meet Australian requirements. All or parts of these ships have been built in Australia. Ship support activities for repairs and modernization have been accomplished in-country by Australian public- or private-sector organizations. The volume of business placed with Australia's shipbuilding industrial base has been sporadic; the consequential employment peaks and troughs have inhibited the development and retention of workers in crucial skill areas and resulted in a less-than-efficient industrial base.[1]

---

[1] For a history of Australia's shipbuilding and ship repair industries and an overview of its current facilities, refer to Appendix A.

What Makes Naval Surface Ship Production and Sustainment Unique

Naval ship production and naval ship sustainment, while having similarities and employing many of the same sets of skills, are distinct both from each other and from their counterparts in the maritime commercial industry.[2]

The design and construction of naval surface ships are some of the most complex weapon system platform engineering, manufacturing, and integration tasks that a country can undertake.[3] These ships require a sophisticated integration of communication, control, weapon, and sensor systems that must work together as a coherent arrangement. These components and subsystems are made up of multiple electronic, mechanical, firmware, and software systems—often leading-edge developments incorporating less-than-benign levels of technical risk.

Naval warships are highly complex, mobile weapon platforms. They generate basic power for speed and mobility, as well as complex power to energize their communication, sensor, and weapon systems. They accommodate, feed, and husband the ship's company who crew, conduct operations, and maintain and repair the ship 24 hours per day.

Given warships' complexity, manufacturing them requires substantial computer-assisted design, engineering, management, testing, and production resources. Engineers from all appropriate specializations are involved in design and manufacturing. Production requires skilled proficiency in many trades: electricians, boilermakers, welders, painters, and so on. Testing complex systems requires commissioning

---

[2] For further discussion of these distinctions, see John Birkler, Denis Rushworth, James Chiesa, Hans Pung, Mark V. Arena, and John F. Schank, *Differences Between Military and Commercial Shipbuilding: Implications for the United Kingdom's Ministry of Defence*, Santa Monica, Calif.: RAND Corporation, MG-236-MOD, 2005.

[3] For the purpose of discussion here, *naval surface ships* include ships that are blue-water capable or that fulfill coastal protection roles.

and test specialists to verify functionality. The workforce for the production trades might peak in the thousands for a typical naval vessel.

Shipbuilding is capital and labor intensive, warship manufacture particularly so, requiring fabrication facilities (to make component parts and structures), docks, slipways, piers, and cranes for assembly and integration activities. These facilities take up large areas of land, must be water accessible, and are expensive to build and maintain.[4]

Naval production relies on a significant and highly diverse contractor supply base that ranges from service support to supply of weapon systems material and equipment. Nations that maintain significant navies (particularly ones with expeditionary capabilities) have established domestic industries that specialize in warship system production and support.[5]

There are many reasons given for why a domestic naval shipbuilding capability might be desirable.[6] Some primary reasons for sustaining such an industry are to accomplish the following:

- Sustain sovereign, national support for sensitive technologies not available from third-party international sources.
- Tailor warships to meet specific national needs or operating doctrine.

---

[4] The shipbuilding industrial base also comprises suppliers of major components: diesel engines, reduction gears, propellers, shafts, valves, actuators, heating and air-conditioning units, switchboards, and so on. RAND's research charter was to focus on shipyard capabilities and labor rather than on the supplier elements of the industrial base. We acknowledge, however, that there are many Australian suppliers that depend on a healthy industry. Further research into the robustness and health of the supplier base was beyond the scope of our charter.

[5] Daniel Todd and Michael Lindberg, *Navies and Shipbuilding Industries: The Strained Symbiosis*, Westport, Conn.: Praeger Publishers, 1996. However, in the maritime commercial shipbuilding industry, a small number of countries dominate the production in a particular market segment (such as cruise ships or bulk containers); see Birkler et al., 2005.

[6] Todd and Lindberg, 1996.

- Control sensitive information about the technologies and capabilities of its warships.
- Address political sensitivities. Most shipbuilding industries employ large numbers of workers, hence politicians have a vested interest in preserving a healthy domestic industry. Many national shipyards (e.g., Navantia, which has built warships in Spain since the Spanish Armada in the 16th century) are long-established firms with extensive histories. National pride makes them even more politically difficult to close.
- Stimulate the domestic economy writ large, given that warships are part of the national defense insurance paid with public funds.

Australia's Naval Shipbuilding Industrial Base

Four main companies make up Australia's naval shipbuilding industrial base today: ASC Pty Ltd, with headquarters in Adelaide, South Australia; Austal, with headquarters in Perth, Western Australia; BAE Systems Australia, a subsidiary of BAE Systems plc, with headquarters in Adelaide, South Australia; and Forgacs, with headquarters in Newcastle, New South Wales. Table 2.1 shows the shipbuilding programs that the companies are currently working on and the locations of their facilities. Figure 2.1 plots the geographic locations of their shipbuilding and repair yards, along with other major yards that are involved with Australian naval shipbuilding and ship repair.[7]

As we discuss in Chapter Three, two warship-building programs are active in Australia today: the *Canberra*-class LHD program, procuring two ships, and the *Hobart*-class AWD, producing three ships. Navantia, a Spanish state-owned company that belongs to the Sociedad Estatal de Participaciones Industriales, built the hull, mechanical, and electrical (HM&E) equipment for the LHD program in Spain to the point where those elements were approximately 65–75 percent complete. Navantia then shipped the basic hulls to the BAE shipyard in

---

[7] For further details regarding Australia's shipbuilding history and current production facilities, see Appendix A.

Table 2.1
Australia's Shipbuilding Industrial Base

Company	Shipyards Currently in Use	Current AUS DoD Shipbuilding Programs	Other Current Shipbuilding Programs
ASC	ASC South Adelaide, South Australia[a]	*Hobart*-class AWD (block construction and integration)	None
	South Australian government's Common User Facility (CUF)[b]		
Austal	Austal Henderson, Western Australia[c]	None	Customs and Border Protection Services patrol boats (PBs)
BAE Systems Australia	BAE Williamstown, Victoria[d]	*Hobart*-class AWD (block construction)	None
		Canberra-class LHD (system installation and final outfitting)	
Forgacs	Tomago, New South Wales[e]	*Hobart*-class AWD (block construction)	None

[a] Located northwest of Adelaide, ASC South is used for the consolidation of the AWDs and is Australia's largest naval shipbuilding hub, incorporating a critical mass of warship design and construction skills. ASC North is a high-tech submarine maintenance facility containing containing shiplift, docking, and transfer system halls; wharf facilities; warehousing; hardstand area; a dedicated painting and blasting facility; and construction and assembly halls. See ACIL Allen Consulting, *Naval Shipbuilding & Through Life Support, Economic Value to Australia*, ACIL Allen report to Australian Industry Group, December 2013.

[b] The CUF is a national strategic asset owned and operated by the government of South Australia. The facility is spread across eight hectares at the heart of Techport, Australia, northwest of downtown Adelaide. See ACIL Allen Construction, 2013.

[c] Located at Cockburn Sound, south of Perth, the facility integrates naval shipbuilding and ship repair with facilities that also support the oil and gas sectors at the Australian Maritime Complex CUF in Western Australia. This CUF was jointly funded by the federal and Western Australia governments to assist local industry with competing for services to the oil and gas, resources, and marine/defense industries. See ACIL Allen Construction, 2013.

[d] Williamstown is situated on the western shore of Port Phillip Bay. It has two building berths; travelling cranes; a graving dock; transporters with large capacity; halls for module construction, assembly, and blast, paint, and outfit; and fully serviced wharfage. Williamstown received a substantial capital upgrade to assist with the AWD project. See ACIL Allen Construction, 2013.

[e] This nine-hectare waterfront shipyard on the Port of Newcastle hosts Forgacs' build of AWD modules. It also is used for marine vessel conversions, refits, unscheduled repairs and maintenance, and survey dockings, with two slipways and extensive wharfage. See ACIL Allen Construction, 2013.

Figure 2.1
Major Australian Shipyards and Their Current Roles in Naval Shipbuilding and Repair

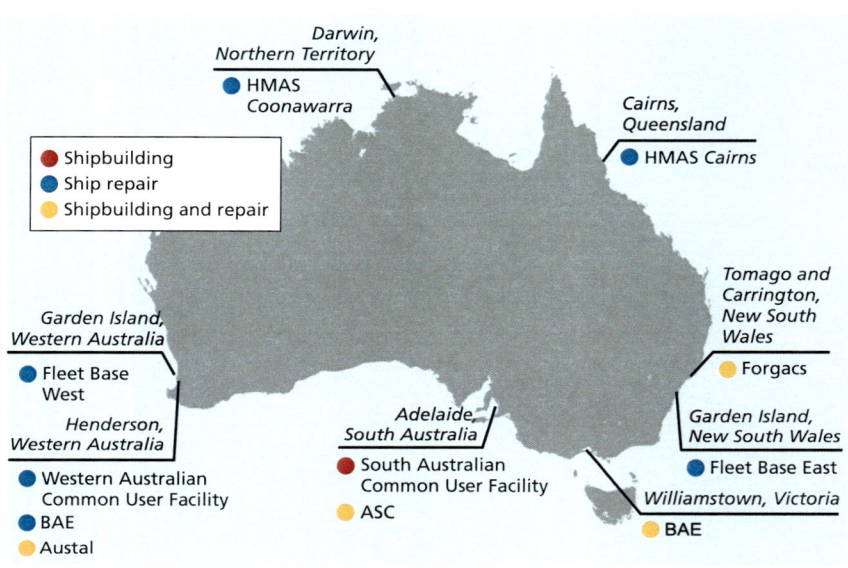

Williamstown, Victoria, where BAE fabricated, equipped, and outfitted the superstructure and completed the final delivery.[8] For the AWD

---

[8] Outfitting tasks occur either during the construction of the pieces that make up the ship or when those pieces are assembled to form the completed ship. Outfitting covers a broad range of functional tasks including:

- structural: installing equipment foundations, doors, ladders, hatches, and windows
- piping: installing and welding pipes, including spools and connectors
- electrical power distribution: installing the power distribution system downstream of the main power switchboards, including hanging and pulling cables and installing local switchboards and ancillary electrical equipment
- Heating, ventilation, and cooling (HVAC): installing air handling units, ducting, and other ancillary HVAC equipment
- joinery: installing accommodations, such as cabins or berths, dining facilities, food preparation areas, and rooms for meetings or other administrative purposes
- painting and insulation: covering the structure and accommodations of the ship.

For naval combatants, outfitting also includes the installation of combat and weapon systems.

program, ASC is the shipbuilder member of the AWD Alliance, with the Defence Materiel Organisation (DMO) and Raytheon Australia Pty Ltd. BAE and Forgacs are allocated blocks through subcontracts, managed by ASC, from the AWD Alliance.[9] ASC supports the AWD program out of its South Australia shipyard, leveraging the South Australian government's CUF. BAE builds AWD blocks out of its Williamstown, Victoria, shipyard, and Forgacs builds blocks at its Tomago shipyard in New South Wales.[10]

Shipbuilding Workforce

The workforce employed in Australia's naval shipbuilding industry has some skills that can cross-locate to the gas, oil, and mining industries. The workforce is shifting and reshaping continually, so a precise analysis of its size and composition has been difficult. Nevertheless, to gain insight into the industry's workforce composition, RAND fielded a survey to the major Australian shipbuilding and ship repair companies and interviewed senior management at each organization. The survey form that RAND used is reproduced in Appendix E.

In its survey, RAND requested information on each company's workforce. We specifically asked for workforce details for the following six broad skill subcategories: general management, technical, structure, outfitting, direct support, and other.[11] Table 2.2 shows the specific skill breakdowns of those skill categories and for all organizations involved with shipbuilding and ship repair (see Appendix B for more detail on the shipbuilding workforce framework).

As discussed in the next section, many ship repair companies use local subcontractors to supplement their workforces. In certain geographic locations, the same subcontractors may work for multiple ship

---

[9] Navantia did support the AWD program by building blocks at the Ferrol, Spain, shipyard.

[10] Austal does not support any current AUS DoD shipbuilding programs but is noted as a shipbuilder in Table 2.2 because of current activity to build patrol boats for the Australian Border Protection and Customs Service in its Henderson, Western Australia, shipyard.

[11] Typically, shipyards analyze their workloads and potential skill shortfalls using much more detailed sets of skills. However, to make data collection feasible, the RAND team aggregated the workforce into a simpler, six-category breakdown of skills. In so doing, we recognize that our analysis might have muted problems with specific skills.

Table 2.2
Taxonomy of Shipbuilding and Ship Repair Skill Categories

Category	Subcategory	Specific Skill
General management and technical	General management	Management
		Administration
		Marketing
		Purchasing
	Technical	Design
		Drafting/computer-aided design (CAD) specialist
		Engineering
		Estimating
		Planning
		Program control/project management
Manufacturing	Structure	Steelworker, plater, boilermaker
		Structure welder
		Shipwright/fitter
		Team leader, foreman, supervisor, progress control (fabrication)
	Outfitting	Electrician, electrical tech, calibrator, instrument tech
		HVAC installer
		Hull insulator
		Joiner, carpenter
		Fiberglass laminator
		Machinist, mechanical fitter/tech, fitter, turner
		Painter, caulker
		Pipe welder
		Piping/machinery insulator
		Sheet metal
		Team leader, foreman, supervisor, progress control (outfitting)
		Weapon systems
	Direct support	Rigger, stager, slingers, crane, and lorry operators
		Service, support, cleaners, trade assistant, ancillary
		Stores, material control
		Quality assurance/control

repair companies. Forgacs acts as a major subcontractor to the providers, and Naval Ship Management (Australia) Pty. Ltd. (NSM), a joint venture between UGL Limited and the Babcock International Group PLC, employs several thousand subcontractors a year, many of whom provide maintenance support for the *Anzac* fleet.

Ship Support Industrial Base

The previous section concentrated on the industrial base that builds new ships, but the companies that provide maintenance and modernization support for the Royal Australian Navy (RAN) fleet are a significant part of the naval ship industrial base. This section describes Australia's in-service ship support industrial base and the potential futures it may face. It first describes the differences in facilities, personnel, and schedules of the new build and support parts of the overall industrial base and how new build and support activities are rarely carried out at the same locations or by the same workforce. We then provide an overview of the current in-service ship support industrial base organizations, including their locations, product lines, and general capabilities. This is followed by a description of how the fleet places demands on this part of the industrial base and how these demands may change in the future. The section concludes by assessing the ability of the current industrial base to meet potential future demands, especially in light of the different paths taken for the shipbuilding industrial base.

Differences Between Shipbuilding and In-Service Ship Support

Building new naval ships, especially major surface combatants, requires specialized facilities and skills. In modern yards, warships are usually built in large structural blocks that are at least partially outfitted. These blocks are built in large, often covered, construction sheds and moved by large capacity cranes for assembly in a dry dock or on a land-level facility. Millions of man-hours are expended over several years, with overlapping demands for hundreds to thousands of skilled personnel. Management and oversight activities closely monitor progress and resolve outstanding issues during the relatively long construction period. The shipyards that build warships may be geographically separated from the ship operating bases.

Ship support includes maintenance and modernization activities, along with engineering, supply, and training support. Maintenance involves periodic tasks that must be accomplished at certain intervals (planned maintenance), as well as the repair and replacement of parts that are defective (operational maintenance). Periodic activities include specific tasks, such as replacing fluids, as well as monitoring various aspects of machinery (e.g., vibration analysis) to estimate when a piece of equipment is likely to fail. Maintenance periods can vary from a few weeks to more than two years, with workforce allocation ranging from a few hundred to a few thousand workers. Demands associated with the material condition of the ship, which are often only discovered during maintenance activities, typically introduce levels of uncertainty that complicate planning, scheduling, and managing such projects. Warship operating cycles over the life of the hull allow fixed periods for extended overhauls; as a result, schedules are tight because the ships are needed back in the fleet to fulfill planned strategic deployments.

Providing support to in-service ships typically requires pier space for ship-berthing during the support period with lower-capacity cranes than those required when building ships. For extended overhauls (major refits) and intermediate dockings, a ship will need to enter a dry dock or land-level facility for more-extensive maintenance, modifications, or routine inspection of propellers, shafting, and underwater appendages. The operating cycle for each class of warship will dictate how often a major overhaul or refit is required; intermediate dockings occur at half that interval. Unplanned emergency dockings occur as required. Most warships will normally require a docking only three to five times during their operational lives. Skilled manpower is needed, but the majority of the required skills are for outfitting or equipment repair.

The longer support periods will typically involve some degree of modernization of a ship's weapon and combat systems. These modernization activities may be provided by the same organization that provides maintenance support or by a separate shipyard that has more-specialized skills. For example, NSM has recently signed a multiyear service agreement to support the maintenance aspects of the *Anzac* fleet, while BAE has responsibility for the major upgrade of *Anzac* systems. The majority of the shipyards that provide support for RAN ships are located very close to the operating bases. The organizations that provide ship repair activities or that accomplish modernizations of naval warships are analogous to those that repair, modernize, or build oil and gas refiners and other facilities.

Because of the different facilities and skills needed to support in-service ships versus those needed to build new ships, there are very few shipyards in the world that accomplish both activities at the same location with the same workforce. As we discuss below, this is true in Australia. This difference in the need for certain facilities and skills is what allows us to treat the two parts of the industrial base separately.

An additional consideration is the fact that shipbuilding involves quantitatively different demands for labor than ship maintenance and modernization. For example, the labor required to build an AWD or assemble an LHD is about three to seven times the labor required for large modernization projects or deep maintenance projects. Moreover, the labor required to build a new ship could be 30 to 70 times more than the labor required for routine maintenance.

Shipbuilding also presents demands for qualitatively different skills than ship maintenance and modernization. In general terms, shipbuilding places greater emphasis on structural work, whereas ship maintenance and modernization work places a greater demand on outfitting. Figure 2.2 shows the distribution of production work (structure, outfitting, and direct production support) for the AWD, LHD, anti-ship missile defense, and two *Collins*-class maintenance dockings. The data clearly show the qualitative differences in demands. As much as 50 percent of the production labor for AWD and LHD was structural, much greater than the 20–30 percent required for ship modernization and maintenance.

For a different view of the same point, Figure 2.3 shows the distribution of labor costs of recent *Anzac*-class maintenance availabilities. The data suggest that just 9 percent of labor costs are related to structure, compared with nearly 38 percent associated with outfitting.

Figure 2.2
Comparison of Production Skills Demanded by Representative Australian Shipbuilding Versus Ship Maintenance Projects

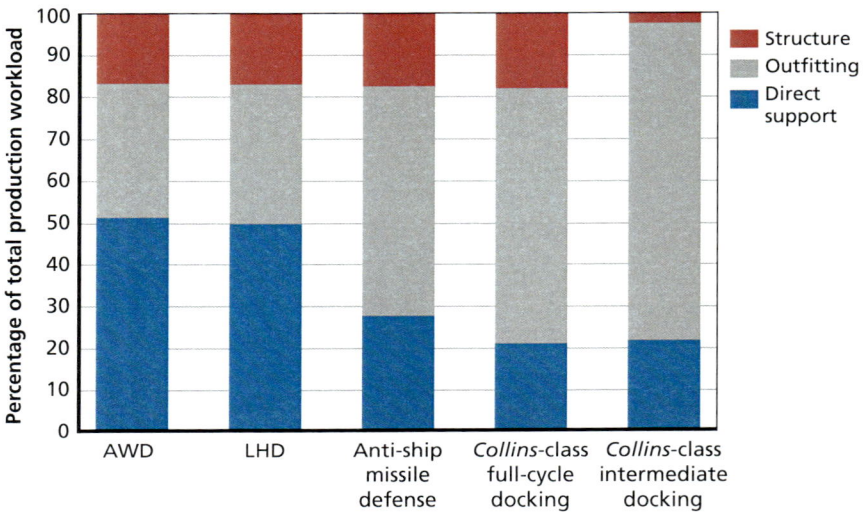

SOURCE: RAND Survey of ASC, BAE, and Forgacs.
NOTE: Data include production work only, excluding recurring and nonrecurring demands for the management and technical workforce. Percentages were taken over the entirety of the program, reflecting both completed and projected demands, as appropriate.
RAND RR1093-2.2

Composition and Capabilities of the Current In-Service Ship Industrial Base

Table 2.3 shows the current organizations and shipyards that provide support to in-service ships. The same and additional companies that are involved in Australia's naval shipbuilding also are involved with naval ship repair. Some organizations—ASC, BAE Systems, and Forgacs—build blocks; perform final assembly, testing, and delivery of new ships; and provide support. Typically, they accomplish support activities at different locations from the new build shipyards. ASC performs longer-term, "deep" maintenance of *Collins*-class submarines out of its shipyard in South Australia and more-routine submarine maintenance at the Western Australian government's CUF; at Fleet Base East, Garden Island, New South Wales; and at Fleet Base West, Garden

Figure 2.3
Percent Distribution of *Anzac* Ship Alliance Labor Costs for Recent Maintenance Projects

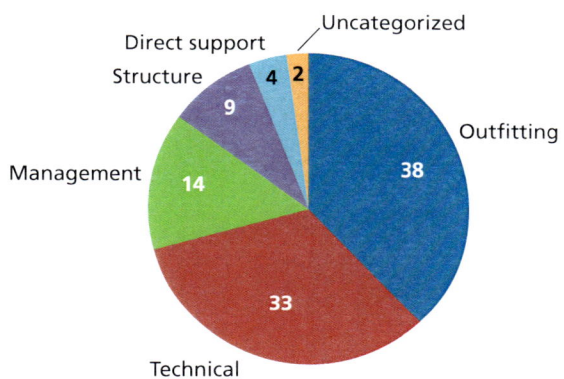

SOURCE: RAND Analysis of *Anzac* Ship Alliance data.
NOTE: Data include HMAS *Stuart* Capability Insertion Opportunity, HMAS *Perth* DSRA04, HMAS *Ballarat* SRA05, HMAS *Parramatta* DSRA06, HMAS *Toowoomba* DSRA04, HMAS *Warramunga* SRA07, and HMAS *Anzac* DSRA10 maintenance projects.

Island, Western Australia. BAE Systems is modernizing the *Anzac*-class frigates at its Western Australia shipyard, and NSM is responsible for maintenance. BAE also is responsible for maintaining Australia's hydrographic ships out of HMAS *Cairns* in Cairns, Queensland, and was recently awarded the contract to sustain *Canberra*-class LHDs. Defence Maritime Systems is responsible for maintenance of *Armidale*-class patrol boats at HMAS *Coonawarra*, Darwin, Northern Territory, and HMAS *Cairns*, in Cairns, Queensland. Forgacs is responsible for maintaining Australia's replenishment ships, the *Tobruk*-class landing ship heavy (LSH) and survey ships out of its ship bases in Carrington and Fleet Base East at Garden Island, New South Wales. NSM and Thales have additional responsibilities for ship maintenance that have not been fully characterized as of this writing. Thales, for example, is responsible for maintaining replenishment and survey ships. Finally, we note that ship repair contracts for the AWD program have not been settled as of this writing.

Table 2.3
Australia's Ship Repair Industrial Base

Company	Shipyards Currently in Use	Current AUS DoD Ship Repair Programs
ASC	• ASC North, Adelaide, South Australia • Australian Marine Complex CUF, Henderson, Western Australia • Fleet Base West, Garden Island, Western Australia • Fleet Base East, Garden Island, New South Wales	• *Collins*-class submarine
Austal	• Henderson, Western Australia • Darwin Naval Base	• *Armidale*-class patrol boats subcontractor to Defense Maritime Systems
BAE Systems Australia	• Henderson, Western Australia	• *Anzac*-class frigate modernizations
	• HMAS *Cairns*, Cairns, Queensland	• Hydrographic ships
	• HMAS *Waterhen*, New South Wales	• Mine hunter coastal
		• *Canberra*-class LHD
	• Williamstown, Victoria	• *Anzac*-class frigate modernizations
Defence Maritime Systems[a]	• HMAS *Coonawarra*, Darwin, Northern Territory • HMAS *Cairns*, Cairns, Queensland	• *Armidale*-class • Patrol boats
Forgacs	• Carrington, New South Wales • Fleet Base East, Garden Island • New South Wales	• Replenishment ships • *Tobruk*-class LSH
NSM	• Henderson, Western Australia • Fleet Base East, Garden Island, New South Wales • HMAS *Cairns*, Cairns, Queensland	• *Anzac*-class frigate maintenance and support
Thales	• Fleet Base East, Garden Island, New South Wales	• *Anzac*-class frigate • Combat system support • Guided missile frigate (FFG) support and upgrades • *Tobruk*-class LSH

[a] Defense Maritime Systems manages repair of survey ships at HMAS *Coonawarra*, Darwin, Northern Territory, and HMAS *Cairns*, Cairns, Queensland.

To remind readers, the discussion earlier in this chapter summarizes what is known about the number of employees at the main ship repair organization; note again that survey data did not discriminate what employees were involved with ship repair as distinct from shipbuilding. Data provided by BAE Systems for its Henderson and Rockingham shipyards do offer some insights. Although these shipyards do some offshore work, they are primarily involved with ship support. Of the 500-plus employees at the Henderson and Rockingham shipyards, approximately 10 percent are subcontractors or contingency workers. The majority of these subcontractors are used in the support of in-service ships. Interviews with BAE indicated that a significant fraction of in-service support engineering is based in Williamstown, adding to these numbers. Interviews with other ship support organizations suggest that the percentage of subcontractors used in the repair and maintenance processes is quite large. Often, subcontractors will support more than one organization, shifting from one project to another as work ebbs and flows. Given the cyclical nature of ship support work, it is typically more economical for ship support organizations to utilize subcontractors than to have permanent employees.

DMO's Maritime Systems Division has started to award long-term support contracts to various companies. For example, a group maintenance contract was recently awarded to NSM (the UGL-Babcock joint venture) for maintenance support for the *Anzac* fleet. Thales also was awarded a contract to support the FFG ships. The support contractors establish man-hour rates for specific tasks, such as removing, repairing, and replacing certain ship equipment. The support organization has responsibility for capacity planning and establishing necessary supply chains. These long-term contracts place more responsibility with the support contractor but provide some degree of long-term stability. And as mentioned, because of the differences between maintenance and modernization workloads, one company may have maintenance responsibilities while another has modernization tasks. This is the case with the *Anzac* fleet, where NSM has maintenance responsibilities and BAE Systems is assigned the current modernization work.

How the Fleet Places Demands on the Support Industrial Base

Each class of ships has a usage upkeep cycle plan. This plan specifies when specific support periods will occur and how long those periods should last. Although the duration of a maintenance period is specified, the future workforce demands during those periods have large uncertainties. There are planned periodic maintenance actions, which typically have established workloads. But it is difficult to predict when repairs are needed and what unexpected, emergent issues may arise, especially as a ship ages. As the start of a maintenance period approaches, reports from the ship's crew and material inspections help to define what tasks need to be accomplished. The support organizations then must plan for the needed spare parts, personnel, and facilities required to support the forthcoming maintenance period. Adding to this uncertainty in future workforce demand is the difficulty in predicting when and what types of modernizations will be needed over a ship's life. Modernization activities may have to be scheduled for a maintenance period sufficient in duration to install the new systems and equipment.

As with workforce demands, the actual start of a maintenance period for a specific ship has uncertainties. At times, fleet operational demands may preclude a ship being available when the maintenance period is scheduled. The overall class maintenance plan typically allows some deviation in scheduling by using upper and lower bounds. These changes to planned schedules can have the effect of having multiple ships in a corrective maintenance period at the same time. Figure 2.4 shows the number of ships in the *Adelaide*, *Anzac*, and AWD classes that are scheduled to be in some corrective maintenance or modernization period on a monthly basis from 2014 to 2019. Before the AWDs enter the fleet, up to six or seven of the 11-ship *Adelaide* and *Anzac* class may be in a corrective maintenance period at the same time. Careful schedule planning is needed to balance operational needs and support industrial base capabilities.

Specific ships in a new class are normally added to the fleet over several years. Planning for this allows the maintenance periods across the class to be synchronized in a way that minimizes large fluctuations in demands over time. In theory, this interweaving of individual ship

Figure 2.4
Number of Surface Combatants in Maintenance, 2014–2019

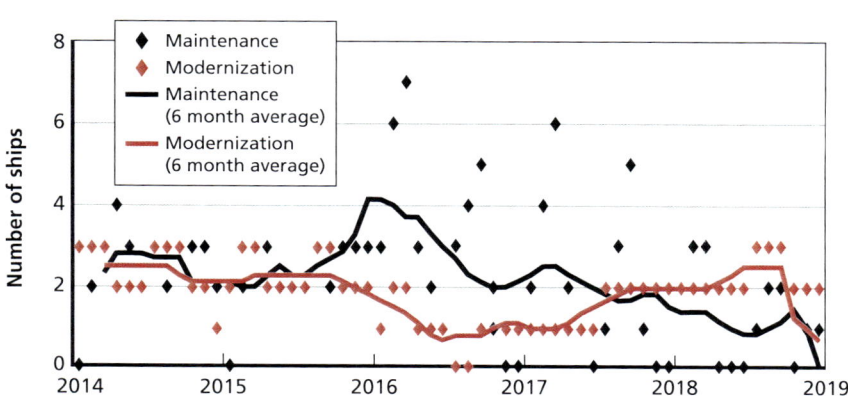

SOURCE: RAND analysis of the *Anzac* System Program Office Ship Maintenance Availability Master Plan, Version 81, July 2014, provided to RAND.
NOTES: The maintenance and modernization data are not necessarily additive. The surface combatants included in this figure are guided missile frigates, frigate helicopters, and air warfare destroyers.
RAND *RR1093-2.4*

maintenance periods across the class facilitates better planning and helps lower costs. However, because of operational needs and unexpected events, maintenance periods can overlap, typically calling for overtime work or additional workforce.

Future Demands for Support

The current in-service ship support industrial base services the current fleet. New classes of ships are under construction, with additional classes planned for the future. These new ships are enhancing the fleet while replacing current classes that are reaching the end of their operational lives. Table 2.4 shows the current composition of the RAN fleet.

The patrol boats and mine hunters are not complex ships. They are supported by an experienced set of subcontractors that also support commercial and privately owned boats. The numbers of these smaller ships are not scheduled to change, but a new class of patrol boats will

Table 2.4
Current Royal Australian Navy Force Structure

Type of Ship	Class	Number	Commissioning Dates
Combatant			
Guided missile frigate	*Adelaide*	4	1983–1984, 1992–1993
Frigate helicopter	*Anzac*	8	1996, 1998, 2001–2006
Patrol boat	*Armidale*	14	2005–2008
Mine hunter coastal	*Huon*	6	1999–2003
Landing ship heavy	*Tobruk*	1	1981
Landing ship dock (LSD)	*Choules*	1	2011[a]
Noncombatant			
Hydrographic ship	*Leeuwin*	2	1997, 2000
Survey motor launches	*Paluma*	4	1989–1990
Support ship		2	1986, 2006

[a] Formerly the Royal Navy *Largs Bay*, commissioned in 2006.

replace the current *Armidale* class. It is anticipated that the new class will have similar or perhaps lesser support demands than the *Armidale* class, at least over the first several years of the operational life of the new class. The industrial base that provides support to the patrol boats and mine hunters is not expected to experience any difficulties meeting future demands. Some noncombatants (the hydrographic ships and survey motor launches) and the HMAS *Choules* will remain in the future fleet, and the organizations that support those ships should not see much change in near- and medium-term workload, although over the long term, it is likely that the *Choules* will require increasing support. Two new support ships will be added to the fleet to replace existing support ships and will eventually require future workload. However, these ships are largely commercial by design and have few military systems. They can be supported at any facility that supports the maritime commercial shipping industry. The two new support ships will not place significant new demands on the support industrial base.

Although segments of the ship support industrial base should not experience any major changes in future workforce demands, there are new classes of ships that will affect future support workforce demand.

The AWDs are new to the fleet. In the context of crewing, they can be considered as replacements for the *Adelaide*-class FFGs. They will be one of the largest and most complex warships operated by RAN. There will be new maintenance demands, especially in the information technology areas, and future modernizations of combat and weapon systems. The new *Canberra* class replaces the former amphibious capability (*Manoora, Kanimbla*, and *Tobruk*). The new LHD-class ships will be larger and more capable than the ships they replace, and there are likely to be different workforce demands (noting that those vessels being superseded had a high maintenance overhead). Following the AWDs will be the Future Frigate. Although those new ships will replace the *Anzac* class on basically a one-for-one basis, they are likely to be larger and more complex than the *Anzacs*. Again, workload may increase and demand different skills. Finally, the future offshore patrol vessels (OPVs) are proposed to replace the patrol boats in due course.

Unfortunately, data are not available on the future support workforce demands from these new classes of ships. The schedule of planned maintenance periods for the new AWDs has been developed, and it is similar to maintenance patterns across other classes of RAN ships. The Future Frigate program is still in the concept analysis stage, and no decision has yet been made about which design will best meet the operational requirements. Whichever acquisition path is chosen for the Future Frigate, the periodic maintenance schedule for the new ships will likely mirror the schedules of other RAN surface combatant ship classes. The periodic maintenance schedules for a specific ship interweave with the scheduled maintenance periods of the other ships in the class. With careful planning and scheduling, workforce demands across the class can be synchronized to minimize large deviation from an average workload. Leveling workforce demand over time with fairly firm demand periods allows the in-service ship industrial base to plan for meeting the demand profile in a cost-effective manner.

In addition to the absence of demand data for the new classes of ships, scant data were received from the surveys on the historical and current workforce demands for ships in the existing RAN fleet. Without data, it is almost impossible to predict future demands on the in-service ship support industrial base. We can assume that the new classes of ships will affect the age of the fleet (see Figure 2.5), and in general, the maintenance demands from newer ships should be less than the demands from older ships. But, as mentioned, the newer ships are larger and more complex than the ships they replace. We estimate that demand will increase slightly but there will be some ups and downs in the demand profile, depending on when new ships come in and old ships go out.

Ability to Meet Future Demands for Ship Repair

As discussed, we received limited data on the current workforce resources in the support industrial base. Some companies did not respond to the survey, and some survey responses combined new build and support workforce at a shipyard. Based on our interviews with the various support organizations and the limited data available, we are left with the impression that the support industrial base is fairly robust and has the ability to expand the workforce to meet future demands. The *Anzac* anti-ship missile defense modernization program will be ending

Figure 2.5
Average Age of Royal Australian Navy Fleet, 2014–2046

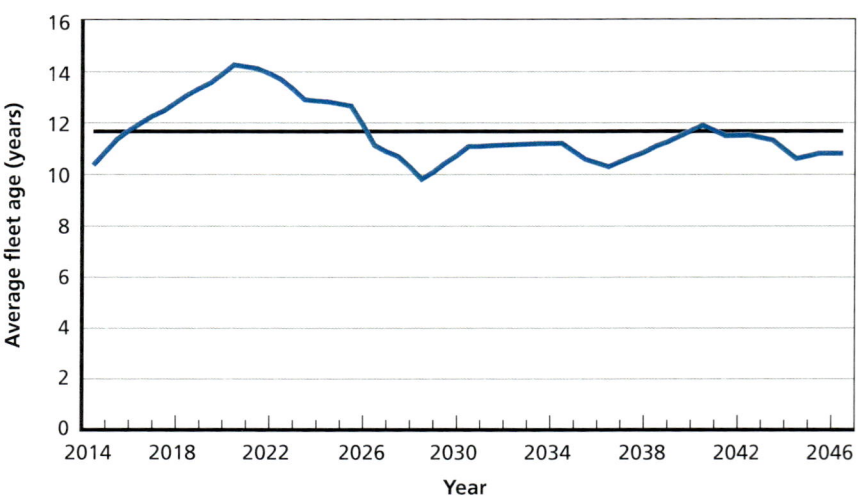

RAND RR1093-2.5

as the first AWD demands start to appear. Workforce should be able to transition from the *Anzac* work to meet the emerging demands of the AWD and Future Frigate, although that work is widely disbursed across the Commonwealth. Also, the workforce demands from building the AWD ships are decreasing and will end at the three shipyards in the next one to four years. Based on decisions for the Future Frigate program, there may be a gap of several years in the demand for new shipbuilding. Construction or repair workforces that are surplus in one area may be available to support respective workforce expansions in another area on a fly-in/fly-out basis.

There are other economic factors that affect workforce availability in the in-service ship support industrial base. Ship repair work is very industrial in nature and has some degree of risk. New workers may decide that shipbuilding or ship repair is not their preferred profession. Other industries, such as construction, mining, and offshore oil and gas extraction may offer higher salaries, especially for workers with proficiency in a skill. If those other industries are expanding, they may draw from the skilled labor that the support organizations use.

Although our initial assessment is that the organizations of the Australian in-service ship industrial base are robust enough to meet future demands, additional data would help to better understand the challenges that the support organizations may face from future decisions on the Australian shipbuilding industrial base.

Observations on the Shipbuilding and Ship Support Industrial Bases

Several broad observations can be made based on this snapshot of the Australian shipbuilding and ship repair industrial bases. First, maintenance and production demands are distributed across a relatively large number of organizations; seven organizations (not including subcontractors) constitute the Australian shipbuilding and ship repair industrial base today. Three organizations build ships, and all seven are involved in ship maintenance and modernization. It is important for the study to answer the following questions: How many shipyards can be supported by the planned shipbuilding program and force structure, and in particular, what are the benefits, costs, and risks of consolidating build or maintenance activities at fewer yards?

A second observation is that Australia's maintenance and production activities are distributed across similar companies, but different shipyards. All the organizations that have a role in production also have maintenance contracts. However, in general, the shipyards that support shipbuilding do not support maintenance and repair activity. An exception to this rule is Austal's Henderson shipyard, which is building and supporting *Cape*-class patrol boats for the Australian Customs and Border Protection Service. A second exception is Forgacs, which builds AWD blocks and has maintenance contracts, but this work appears to be managed at different facilities within Forgacs. This is not unusual, because ship repair and shipbuilding place demands on different skills and require different facilities. Moreover, because ship *repair* is largely predictable and periodic (and is a direct consequence of force structure), those demands will remain essentially constant under the shipbuilding scenarios being examined in this study. *Shipbuilding* demands are highly variable and are a consequence of budget and national security decisions.

Third, although Australia's warship-building workforce headcounts do not include subcontractors, the overall numbers by category appear small in comparison with other large shipbuilding projects conducted overseas. A question for the study is how the current workforce compares with the levels needed to sustain the shipbuilding industrial base. The industrial base modeling tasks in this study will more formally assess supply relative to potential future demands.

A final observation is that Australia nonetheless has a workforce of several thousand workers with experience directly relevant to shipbuilding. About 4,000 of them are currently involved in shipbuilding. The risk of a gap in ship construction is that this workforce will leave the industry and then need to be rehired and retrained once decisions are made to begin producing new ships, such as the Future Frigate.

Australia's Unique Market Niches: Warship and Submarine Production and Support

Successful commercial and military shipyards typically have niche product lines. These yards concentrate on either military or commercial ships and usually specialize in a specific ship type. For example, commercial yards specialize in liquid bulk carriers, dry bulk carriers, container ships, specialized vessels, cruise ships, ferries, and so on, while yards building military vessels focus on specialized complex ships that are produced in limited numbers, such as small patrol boats, frigates, destroyers, amphibious, auxiliary, submarines, and aircraft carriers. In general, each of these niches is characterized by a high degree of specialization, is complex and knowledge based, has tailored production processes and facilities, and is coupled with considerable technical expertise and a high number of specialized subcontractors.

In addition, a shipyard's niche is further defined by what its shops, docks, cranes, piers, and space can handle in terms of length, beam, or draft. Constructing and assembling ships larger than its facilities can accommodate is not possible. But building ships smaller than the shipyard was designed to construct is not efficient either, because the yard is not fully utilized; technical, workforce, and subcontractor expertise is not as deep, and overhead costs are spread over a smaller business base.

It is very difficult to work on both military and commercial ships, because military specifications are almost always more stringent and can require extensive engineering capabilities and worker skill levels that are unnecessary and too expensive for commercial work. Moreover, commercial work is focused on timely throughput, because most commercial ships have significant penalties for being late. This schedule adherence is, in part, achieved through minimal changes once production begins. Thus, it is rare for a commercial ship to be late.

In military work, submarines are a different product line because of even more stringent specification requirements and submarine safety processes and inspections. Also, tooling and fixtures are quite different due to the differences in hull shapes and materials. The system density for submarines makes outfitting much more challenging and requires trades to have specific training. The one shipbuilder in the United States that manages both product lines generally segregates its construction workforce between submarines and surface ships (carriers).

Both military- and commercial-focused shipyards also tend to specialize in either repair and modernization of existing ships or construction of new ships. New construction requires a heavy emphasis on fabrication skills (e.g., steel fabrication and erection) and production facilities (e.g., block assembly, steel panel lines). Repair and modernization shipyards require a greater proportion of outfitting skills than fabrication. Also, repair facilities require more machine and electrical shop capacity.

Aside from worker skills and facilities, each type of work offers significantly different management challenges. New construction is a far more structured process than repair, with a focus on material acquisition and orderly module fabrication, erection, outfitting, testing, and trials. Repair starts with inspections, equipment removal, repair planning, and repair part procurement. It must adapt to unknown conditions and problems that are only discovered once work begins. As a result, repair work processes are designed to be more flexible to emergent work compared with new construction. After equipment is refurbished (in hull or off) or replaced, it then must be installed on and reconnected to the ship, whereupon it needs to be tested, aligned, and integrated.

Australia's principal warship construction shipyard now has two product lines (and the potential for a third with new submarine construction). Currently, ASC performs heavy maintenance for the *Collins*-class submarines at Osborne and Henderson and is constructing the AWD at Osborne. This diversity of work results from episodic demands for warship construction and from the Commonwealth's desire that RAN ships be built in domestic shipyards.

As the Commonwealth embarks on modernizing its naval forces and creating a naval shipbuilding enterprise, it may well want to consider alternative industrial structures. If it were to segregate surface combatant from submarine construction and assembly facilities and workforces and to similarly separate construction and assembly from repair and modernization, Australia could enable domestic shipbuilders to concentrate on specific value-added niches and avoid relying on a single organization to manage multiple, disparate product lines simultaneously.

CHAPTER THREE
Australian Department of Defence's Planned and Projected Warship Acquisitions

In this chapter, we explore the acquisition outlook that Australia's naval shipbuilding and ship repair industries will face in the next several decades. This outlook portends significantly different issues for each industrial base. For shipbuilders, the main issue is whether it is possible to sustain new shipbuilding in Australia in a cost-effective manner. Shipbuilders are wholly dependent on new acquisition programs. Companies involved in ship repair depend upon the size and composition of the RAN fleet for their livelihoods. As described in Chapter Two, as long as the fleet stays fairly constant in size (or increases) and the ships are supported in Australia, the focus of the ship repair industry is how best to organize and utilize its various ship repair resources to their best commercial advantage.

Current Shipbuilding Programs

Australia has two active shipbuilding projects for RAN: the landing helicopter dock and the air warfare destroyer.

Landing Helicopter Dock Project

The *Canberra*-class LHD project is a two-ship class being built for RAN. It is part of a project to upgrade Australia's amphibious capabilities, based in part on lessons learned from the East Timor peacekeeping operation. Construction of the first ship, the HMAS *Canberra*, began in 2008, and the hull was launched in 2011. Work on the second ship, the HMAS *Adelaide*, began in 2010, and it was launched in July 2012.

At about 230 m in length and a displacement of 27,500 metric tons, the LHD is the largest vessel in RAN. In spite of its size, the draft of the ship is only about 7 m, which was a critical design specification given the desire to operate the ship in littoral waters. The LHD will be able to transport 1,046 soldiers and their equipment and can deploy reinforced companies of 220 soldiers at a time. It has two vehicle decks and can carry tanks and armored vehicles. Its well deck holds up to four landing craft. The flight deck can accommodate four helicopters of *Chinook* size or six smaller ones. Current plans call for the ships to homeport in Fleet Base East in Sydney. The Spanish firm Navantia is responsible for constructing the ships, and BAE Systems Australia is fabricating their superstructures and equipping the ships.[1]

Air Warfare Destroyer Project

The *Hobart* class is a derivation of Navantia's F100 design—an enhanced replacement for the *Adelaide*-class FFG-7 frigates. Currently, plans are for three ships.[2] Its primary role is air defense, but it can also fill anti-surface, antisubmarine, and naval gunfire roles. The AWD Alliance has oversight of the building program. Construction involves assembling 31 prefabricated modules, of which nine are being manufactured by ASC in Australia and the others are being contracted to other firms. Schedule problems occurred when a central keel block manufactured by BAE Australia was found to be incompatible with other modules. In 2011, the government announced a two-year delay in completing the ship. The *Hobart's* keel was laid in September 2012 and the *Brisbane's* in 2014. Originally, the new destroyers were to be operational between 2014 and 2017, but those dates have been slipped to 2016 and 2019.[3]

---

[1] This description was drawn from Naval Technology, "Canberra Class Landing Helicopter Docks (LHDs), Australia," web page, undated; IHS, *Jane's Fighting Ships* (online), undated; and other publicly available sources.

[2] There had been a contractual option for a fourth AWD, but that has expired. Our analysis in subsequent chapters assumes that if that fourth AWD were built, it would be the same design as the previous three AWDs.

[3] This description was drawn from IHS (undated) and other publicly available sources.

Future Shipbuilding Programs

In addition to the current LHD and AWD projects, Australia has a robust shipbuilding acquisition program with new projects in various stages of gestation.

Future Submarine Project

Australia remains committed to acquiring a regionally superior conventional submarine that can meet the nation's key requirements for range, endurance, payload, stealth, and sensor performance, avoiding a capability gap at the withdrawal of the *Collins* class. Through an acquisition strategy announced by the government on February 20, 2015, AUS DoD will employ a competitive evaluation to select an international partner to develop and deliver the future submarine. France, Germany, and Japan will be invited to participate in the competitive evaluation, which will inform a decision on the international partner for Australia's future submarine.

Future Frigate Project

The Future Frigate will complement the AWDs described above and replace the *Anzac*-class frigates, which will be retired progressively starting in approximately 2026. The Future Frigate program is in the early stages, with various options still being examined for the preferred acquisition path. One option is to modify the AWD hull form (basically, the HM&E) to meet the Future Frigate desired operational requirements.[4] A second option is to modify an existing MOTS design.

---

[4] These modifications would include replacing the Aegis system with a CEAFAR radar and a Saab combat management system (the systems currently being installed on the *Anzac* class during their major modernization). They might also include changing the base design to support two helicopters (the AWD currently supports only one helicopter) and to address any outstanding environmental issues. Also, the AWD design may require some modifications to enhance the ability to perform the primary antisubmarine warfare mission. These design modifications will take some time once contracts are signed and, as suggested by the AWD and other programs, construction should not begin until the majority of the final drawings for construction are complete.

It is likely that any specific MOTS design will require similar design modifications as those needed to adapt the basic AWD design.[5]

Given that it is unclear when there will be a final decision on which acquisition path—common AWD hull or evolved MOTS—will be chosen for the Future Frigate program, there is some uncertainty of when construction could start for the frigates. The construction and delivery of a first-of-class ship of the size and complexity of the Future Frigate could take five to seven years—the shorter time if the common AWD hull is chosen, the longer time if a new ship design is being built. With the desired in-service date of 2026, construction should start in the 2020 to 2022 time frame.

Supply Ship Project

The government intends to replace the capability currently provided by the supply ships HMAS *Success* and HMAS *Sirius* as soon as it can. Originally, this was to include examining options for local, hybrid, and overseas builds or for leasing an existing vessel.[6] However, in June 2014, then–Defence Minister David Johnston announced that a "limited invitation to tender" had been released to Navantia (a Spanish company) and to Daewoo Shipbuilding and Marine Engineering (a Republic of South Korean company) for an overseas build of the two replenishment ships based on existing designs.

Mine Hunter and Hydrographic Ship Replacements

The *2009 Defence White Paper* established a goal of rationalizing patrol, mine warfare, and hydrological survey vessels into a single class of OPVs, or littoral multirole vessels (LMRVs).[7] The proposed concept

---

[5] Again, if this path is taken, it will require some time to evaluate and choose a MOTS design, negotiate a contract with the design and build firm, make the necessary design changes to meet Future Frigate requirements, and develop the design drawings for the specific shipyards that will be involved in constructing the ships. The path could be shortened if the basic MOTS design can meet the desired operational capabilities—for example, if it could support two helicopters and was designed for antisubmarine warfare missions.

[6] Commonwealth of Australia, 2013a, p. 85.

[7] Sometimes these vessels are also referred to as offshore combatant vessels (OCVs). See Commonwealth of Australia, *2009 Defence White Paper: Defending Australia in the Asia*

envisions that, rather than one vessel filling all three roles, a common hull and propulsion system would be fitted with containerized modules for each mission. Modularization would enable any one ship to be reconfigured for another role. If successful, the new ship will replace four other types of ships: the *Armidale*-class patrol boat, the *Huon*-class mine hunter, the *Leeuwin*-class survey vessel, and the *Paluma*-class survey motor launch.[8] The *2013 Defence White Paper* remained committed to the OPVs, but it announced that there would be an interim vessel to replace the *Armidales*, while the *Palumas* and *Huons* would undergo upgrades.[9]

Pacific Patrol Boats

In June 2014, Foreign Minister Julie Bishop and then–Defence Minister Johnston announced a new AUD 2 billion Pacific patrol boat program that would replace the fleet of 22 patrol boats that Australia gifted to 12 Pacific Island countries from 1987 to 1997 (Cook Islands, Federated States of Micronesia, Fiji, Kiribati, Palau, Papua New Guinea, Republic of Marshall Islands, Samoa, Solomon Islands, Tonga, Tuvalu, and Vanuatu). AUS DoD will offer replacement patrol boats to all current participating states, along with Timor-Leste (East Timor), which has been invited to join the program. Under the new Pacific patrol boat program, AUS DoD will undertake an open tender for the procurement of more than 20 steel all-purpose patrol vessels worth AUD 594 million; it also will include an option through life sustainment and crew training, and other personnel costs, estimated at AUD 1.38 billion over 30 years.

As of this writing, Australia is holding discussions with Pacific patrol boat states on the individual allocation of patrol vessels.[10]

Pacific Century—Force 2030, Department of Defence, 2009.

[8] Sean Thornton, "The Rationale for the RAN Offshore Combatant Vessel," *The Navy (Navy League of Australia)*, Vol. 72, No. 1, January 2010.

[9] Commonwealth of Australia, 2013a.

[10] See Commonwealth of Australia, "Minister for Foreign Affairs and Minister for Defence—Maritime Security Strengthened Through Pacific Patrol Boat Program," Canberra, Australia: Australian Department of Defence, June 17, 2014a.

Emerging Short-Term and Longer-Term Demand Gaps

The three current shipbuilders—ASC, BAE, and Forgacs—are rapidly approaching the end of their work on the AWD and LHD programs.[11] Table 3.1 outlines projected completion dates for current production programs based on responses in our survey of companies. Forgacs will finish its AWD blocks around the third quarter of 2015. BAE will finish its LHD work around the third quarter of 2015 and its AWD work in the second quarter of 2016, although that AWD work will start to decrease significantly about a year before that. ASC's structural and outfitting work on the AWD program will begin to decrease in 2017 and will be completely finished in 2019. The three shipbuilders have already started to shed workforce and, barring new programs, they will have no more structural or outfitting work after the second quarter of 2019.

The current ship acquisition plan reveals that the government's planned acquisition strategy contains a short-term gap in demand for naval ships that will arise with the end of the AWD production and

Table 3.1
Projected Work Completion Dates for Current Production Programs

Program	Company	Projected Date of Work Completion (All Skills)[a]
AWD	ASC	2nd quarter 2019
	BAE	2nd quarter 2016[b]
	Forgacs	3rd quarter 2015
LHD	BAE	3rd quarter 2015

[a] These projected completion dates are for all skills. At a more micro level, certain skills may have earlier completion dates. Nonrecurring technical work may continue past these dates at low levels. Where the fidelity of data permits, we report projected completion of structure and outfitting work.

[b] BAE projects that recurring structure and outfitting work will complete in the second quarter of 2016, but the actual level of effort is projected to decrease significantly starting in the fourth quarter of 2015.

---

[11] As mentioned, Austal is currently building patrol boats for the Australian Customs and Border Protection Service, but that work is also nearing completion.

before the start of the Future Frigate program. Further into the future, a longer-term gap will arise when production of the Future Frigate ends, around 2035. The Australian White Paper team provided two acquisition scenarios as options to the current procurement plan. While similar, the scenarios show different numbers of patrol boats that RAN would acquire and different production start dates for offshore patrol and littoral multirole ships. The two scenarios are outlined in Table 3.2.

Both scenarios would see the acquisition of eight to ten Future Frigates to replace the *Anzac* class, with in-service dates between 2026 and 2035, and three *Hobart*-class destroyers to replace the *Adelaide* class, with in-service dates between 2016 and 2019. The scenarios differ in the number of new patrol boats to replace the *Armidale* class that would come into service between 2021 and 2026: 14 patrol boats in Scenario 1, and six to eight in Scenario 2. The scenarios also differ in the date for the OPVs/LMRVs to enter into service: 2035 in Scenario 1 and 2026 in Scenario 2.[12]

Given the various potential ship acquisition futures and the current state of the shipbuilding industry, the Australian government faces major decisions on the future of Australia's naval shipbuilding enterprise. There are various strategies or paths, ranging from building naval ships in Australia to buying them as turnkey systems from foreign shipbuilders. Australia has faced these questions in previous naval ship acquisition programs and has used various strategies.

Table 3.2
Two Alternative Royal Australian Navy Acquisition Scenarios

	Ship Type Number and In-Service Date			
Scenario	Future Frigate	*Hobart*	Patrol Boat	Offshore Patrol/ Littoral Multirole Vessel
1	8–10 (2026–2035)	3 (2016–2019)	14 (2021–2026)	21 (2035–??)
2	8–10 (2026–2035)	3 (2016–2019)	6–8 (2021–2026)	21 (2026–??)

---

[12] Scenarios were used for the purpose of modeling Australia's future demand profile. They are subject to change based on the final outcomes of the Force Structure Review and White Paper process.

Alternative Strategies for Australian Naval Ship Acquisitions

While there are numerous alternative naval ship acquisition paths that Australia could take in the absence of an indigenous design capability, in this analysis, we focus on six warship acquisition strategies that are reasonable and potentially achievable. Each addresses the issues of *how to build* and *what to acquire* in different ways. The strategies are laid out in Table 3.3 and range from acquiring a MOTS design in its entirety to modifying an existing design to accommodate Australian military, statutory, and regulatory requirements. They include the options of building ships entirely in-country, partially in-country, or overseas.

Of the six paths, we believe four are viable options; these are shaded green in Table 3.3. The two others—building a MOTS design in Australia or partially building and outfitting an existing design with-

Table 3.3
Alternative Naval Ship Acquisition Strategies Open to Australia

How to Build	What to Acquire	
	Military-off-the-Shelf	Modified Existing Design
Build in Australia	FFG[a]	*Anzac* AWD *Collins*
Partially build and outfit in Australia		*Canberra*-class LHDs
Build offshore	FFG,[b] RAN afloat support capability ships	Strategy under consideration for the future submarine

[a] RAN presently operates four FFG-7 frigates: FFG 03 through FFG 06. The last two, FFG 05 and FFG 06, were built by AMEC in Williamstown, Victoria. Ship designer Gibbs & Cox provided the design package, and Unisys provided the combat system. Both were provided through a Foreign Military Sales package. The ships were MOTS, based on the U.S. Navy's *Oliver Hazard Perry* FFG class. Australia selected which U.S. Navy ship alterations or ordnance alterations it wanted in the design. Any other modifications were relatively minor. The first ships were designed in the United States at Gibbs & Cox and built at Todd Shipyards in Seattle, Washington (now known as Vigor Shipyards). During the initial phases, Australia located personnel at the shipyards. Later, when production moved to Australia, Todd provided extensive on-site construction and material support (down to fasteners), while Gibbs & Cox assigned an experienced designer at Williamstown for FFG 05 and FFG 06.
[b] FFG 03 and FFG 04 were built at Todd.

out modification—are unlikely future strategies; these are depicted by the gray-shaded cells.[13] Australia has experience with three of the four green-shaded strategies; it has not used the combination of building offshore and modifying an existing design but is considering it for the future submarine.

These are not necessarily mutually exclusive paths. Australia in the past has combined two strategies as a program has matured. As an example, the first tranche of the FFGs was built in the United States; the last two FFGs, which were based on an identical design, were built entirely in Australia.

Modify an Existing Design to Be Built in Australia

The shipbuilding strategy most recently used is to buy an existing design from overseas and build it with modifications in Australia.[14] This approach is being used for the AWD and was used on the *Anzac* frigates and *Collins*-class submarines, but it has many current critics. While the modified Navantia F100 design drawings for AWD accommodate Australian military, statutory, and regulatory requirements, the drawings were not adapted to ensure efficient construction at the

---

[13] Note that Australia has pursued the gray-shaded acquisition strategies in the past. In the late 1980s and early 1990s, two frigates from the FFG-7 class—*Melbourne* (FFG 05) and *Newcastle* (FFG 06)—were built to print by the Australian Marine Engineering Corporation (AMEC) in Williamstown, Victoria. Assuming a MOTS design meets Australia's military, statutory, and regulatory requirements, there is no guarantee that a strategy to build in Australia and use MOTS equipment would be successful. This design would likely be optimized for construction in a shipyard in the overseas country, not Australia. Shipyards all have different material handling, cutting, machining, and other capabilities that can make successful production of a design in one shipyard inefficient and difficult to produce in another. Moreover, material suppliers may also have different capabilities that affect production. If the overseas shipyard has access to larger steel plates from its mill than an Australian shipyard can acquire, that overseas ship can be produced with fewer welding labor hours. Similarly, weight-handling and plate-cutting size restrictions may require smaller plates to be purchased, even if larger plates are available in Australia, requiring more welding hours. If the Australian shipyard has greater physical capabilities than the overseas shipyard for which the design was optimized, the Australian shipyard may still have to build the ship inefficiently to avoid extensive drawing changes.

[14] By *modified existing design*, we mean a current, proven design modified to meet Australian military, statutory, and regulatory requirements.

chosen building shipyard in Australia. The F100 design was optimized for construction at Navantia's Ferrol shipyard in Spain. The drawings did not take into account different handling, cutting, and machining capabilities for Australian-sourced materials in an Australian building yard. These issues were compounded by different work practices and build strategies. Differences in habitability requirements—berthing, medical, dental, HVAC, and so on—are good examples of design issues that need to be considered early in the design adaptation phase.[15]

To improve this process for future projects, DMO would have to pick (by competition or other means) the specific Australian building shipyard and a foreign design entity. DMO would then form a collaborative association. The partners would form a design-build team equipped with appropriate hardware and software to execute transoceanic collaboration. The team would be colocated at the design site with

- shipyard production planners and experienced tradespeople to ensure the design is optimized for the building shipyard
- empowered DMO and RAN representatives to negotiate how Australian military, statutory, and regulatory requirements are to be met
- ship designers.

Clear top-level specifications need to be established at the outset and regular design review sessions held. Most sessions could be held from the design site, with digital data transfer facilities between the design authority, the building shipyard, and DMO. A small team of Australian engineers, designers, and shipyard planners must be located at the design site to ensure an effective, executable design at the Australian building yard. On occasion, the design review should be undertaken at the building shipyard with DMO representatives present and the design team digitally linked. Such an approach ensures not only that Australia's governmental interests are met (military, statutory, and regulatory) but that cost and schedule risks are mitigated by ensuring

---

[15] For example, bunk size needs to accommodate taller RAN sailors, a change that ripples across a ship's design.

that the detailed drawings necessary for production are optimized for efficient and effective execution in the selected Australian shipyard.

Modify an Existing Design to Be Partially Built and Outfitted in Australia

Building the hull of a MOTS ship overseas for subsequent outfitting and integration of Australian habitability, communications, and combat systems in Australia is certainly feasible for some classes of warships with low levels of complexity. This approach is being used for the *Canberra*-class LHDs; the government is buying portions of a MOTS ship (i.e., HM&E) from Spain and building, outfitting, and integrating the remaining elements in Australia. In this case, Navantia built up to the flight deck in Spain, and the hull was then transported to BAE Systems at Williamstown, Victoria. There, workers constructed and outfitted the superstructrure, undertook final system integration, and conducted trials. France used a similar approach with its *Mistral*-class landing platform/docks (LPDs, also called amphibious transport docks), where partial hulls for early ships were built in Poland and then transported to France for integration, outfitting, and trials.[16] The nature of amphibious warfare ships—with large, mostly empty spaces for troop accommodations, cargo storage, and associated facilities for troop and cargo transportation—has shown this approach to be more feasible than would be the case with a more complex and equipment-dense surface combatant.

Surface combatants are sophisticated ships with highly complex levels of integration between communications, multiple sensors, combat systems, weapon systems, signature reduction measures, and human interfaces. Building a surface combatant's HM&E structure overseas and subsequently integrating its weapon and combat systems in Australia would require extensive transoceanic design and planning. Issues with weapon and combat system technology transfer would need to be resolved. The highly integrated nature of a surface combat-

---

[16] See Laurence Smallman, Hanlin Tang, John F. Schank, and Stephanie Pezard, *Shared Modular Build of Warships: How a Shared Build Can Support Future Shipbuilding*, Santa Monica, Calif.: RAND Corporation, TR-852-NAVY, 2011, p. 74.

ant means that the HM&E building shipyard would require detailed information about and thus gain knowledge of the weapon and combat systems to be installed in the Australian yard, inasmuch as the building yard would provide foundations, power, and cooling capabilities for those systems. Cable runs to connect sensor, weapon, and combat systems will likely be more efficiently performed by the overseas building shipyard than the integration shipyard, necessitating disclosure of more technical data. Moreover, the building yard also may have to include weapon-handling equipment in the hull to avoid extensive hull disassembly in Australia. While feasible, the procurement of these sensor, weapon, and combat systems from third-party countries will introduce issues of intellectual property protection and International Traffic in Arms Regulations at increasingly complex levels between the overseas build yard, the Australian build yard, and weapon system suppliers.

Buy a MOTS Ship from Overseas and Use It Without Modifications
This alternative assumes the following:[17]

- Designs for overseas surface combatants or support ships exist that meet Australia's military requirements.
- Australia can define a flexible set of requirements.
- RAN or DMO obtains authority to waive Australian statutory and regulatory requirements not met by overseas surface combatants or support ships.

This alternative recently has worked for Australia in acquisitions of several weapon systems platforms, including C-130, C-17, and F/A-18 aircraft. Previously, RAN acquired FFG 01, 02, 03, and 04 from the United States, where they were built at Todd Shipyards in Seattle, Washington. FFG 05 and FFG 06 were built in Williamstown, Victoria, by AMEC. To use this alternative again on a surface combatant or support ship would require Australia to make value judg-

---

[17] This strategy is currently being employed to replace the existing RAN afloat support capability. As mentioned, a request for tender has been issued to Navantia (Spain) and Daewoo Shipbuilding and Marine Engineering (Republic of South Korea) for an overseas build of two replenishment ships based on existing designs.

ments on requirements, statutes, regulations, economics, and schedule. To succeed with this alternative, RAN and DMO would also need to obtain the data rights and repair parts necessary to service, repair, overhaul, and modify the ship and its systems over the life of the class.[18]

In addition to the approach of building offshore and modifying an existing design, which Australia has not yet used, these three acquisition strategies define possible paths for the future of Australian naval shipbuilding—build complete new ships in Australia, build parts of a new ship overseas with final construction and outfitting in Australia, or buy naval ships from foreign shipbuilders. These strategies need to work with acquisition plans that Australia may implement, which we briefly discuss in the next section. We will analyze the shipbuilding labor implications of these three strategies from both a short-term and a longer-term perspective in Chapter Four.

Australianization Issues for National Ship Programs

As we have noted, Australia possesses limited domestic capability to design warships that are larger than patrol vessels. For the foreseeable future, Australia likely will need to depend on foreign partners to provide preliminary and detailed designs of its sophisticated warships. Any design obtained from overseas will require modifications and tailoring to meet Australian requirements and shipyard capabilities.[19]

This is the model that Australia has employed over the past 20 years. It has produced three major classes of warships by using designs that were procured from overseas ship designers and then building the vessels in Australia. However, each of these programs, to varying degrees, failed to fully develop the complex and integrated relationship that needs to arise between the ship designer and shipbuilder or to appreciate its influence on production costs and schedule.

---

[18] For the FFG-7 class, RAN purchased repair parts from the U.S. Navy supply system.

[19] In this report, we refer to this activity of modifying an existing design to meet Australian operational, habitability, or environmental requirements as *Australianization*.

The AWD program starkly illustrates these challenges.[20] The ship class is currently under construction by an alliance consisting of the Australian government (represented by DMO), ASC, and Raytheon Australia. The AWD's design is largely a copy of the F100 *Aegis* frigate, originally built in Navantia's Spanish shipyard. The contract calls for design and delivery of three ships.

Even though Navantia has been contracted by DMO outside of the alliance arrangements, the alliance is responsible for administering the contract on behalf of DMO, within certain limits.[21] Navantia's scope covers everything up to production design (construction drawings), and its products are delivered as two-dimensional PDF files.

In building Navantia's design, Australian shipyards are encountering significant issues with the design-build process and the relationship between the shipyard and designer.

Design and Build Issues

The AWD is relying on a model in which a third party designs a ship without reference to the contracted shipyards' capabilities, facility constraints, and build plan. However, the relationship between designer and shipyard is highly complex and needs to allow the design to be iteratively developed to ensure that, in addition to achieving the ship's specifications, it is optimized for producibility in the shipyard. Preferably, this should involve a shared three-dimensional integrated CAD design environment in which design responsibilities are progressively transferred to the shipyard during detailed design phases.[22] The shipyard should be responsible for turning three-dimensional CAD data

---

[20] See Australian National Audit Office, *Air Warfare Destroyer Program*, Audit Report No. 22 2013–14, March 6, 2013.

[21] There is no single prime contractor; the AWD Alliance operates on a collaborative basis. However, the two industry participants (ASC and Raytheon Australia) hold joint liability for contract performance and delivery of the ships.

[22] The use of two-dimensional PDFs and the shipyard's limited exposure to the design before production commences have caused significant difficulties in resolving design issues encountered during production (e.g., clashes, missing data, interface mismatches). Without access to three-dimensional CAD models or knowledge of the design, these issues must be referred back to the designer in Spain for resolution. This takes time and delays production.

into construction drawings, thereby allowing it to optimize drawing content and to structure drawing packages to reflect the shipyard's build plan.

Of particular note is the fact that the AWD design process has not provided an environment in which the producibility of the design can be optimized for the local shipyard. There are two reasons for this: The process has been based on an existing design, and the designer has been contracted only to update the design and deliver it as if it were to be built in its own yard.

Relationship Between Shipyard and Designer
The contracting model adopted for the AWD puts the Spanish designer at arm's length from the constructing Australian shipyards; as a result, the designer is not invested in the overall build outcomes. Given the complexity of the relationship between the designer and shipyard, this provides limited opportunity for resolving performance issues.

For earlier programs, including the *Collins* and *Anzac* classes, the designer was subcontracted to the shipyard, and the shipyard was therefore responsible for the design and the ship's performance. If issues arose in the design, the shipyard was able to rectify them and balance commercial and technical issues.

With the AWD, on the other hand, the designer is responsible to DMO for design performance. The AWD Alliance is separately responsible to the DMO for build but has no design responsibility. Under this model, multiple stakeholders are involved, so any decision is significantly more complex and requires the concurrence of all parties, including the designer.

Observations About Adopting Foreign Designs for Australia
As Table 3.3 shows, two of the four acquisition strategies that likely are viable for Australia involve building naval ships in Australia. Regardless of where ship designs originate, experience shows that the designs for lead ships contain errors. That is the nature of how first-of-class

---

Given the large number of design issues found in the AWD, this has significantly impacted production productivity.

naval ships come about. Moreover, Australia often significantly modifies existing designs to such a degree that they become de facto first-of-class efforts, with the attendant challenges to production and design that such vessels encounter. Additionally, designers are not in proximity to the constructors, and the time lost communicating and correcting discrepancies is very expensive.[23] Such design errors, if not caught early in the process, can add significant time and cost to the overall construction schedule and budget—sometimes increasing the schedule length substantially and labor hours by a factor of two, if not more.[24] Moreover, veteran supervisors and construction trade workers are generally not familiar with the ship, with most seeing the design for the first time when building begins in Australia.

This suggests that future programs pursued by Australia should recognize that successful projects require as close a relationship as possible between ship designer and shipbuilder. Programs need to ensure that the designer is incentivized to make the shipbuilder succeed. Australia should also consider the following tactics:

- Source naval ship designs from one or two foreign countries that possess similar naval architecture standards as Australia.
- Select designs based on producibility in Australia.
- Have an Australian shipbuilder or shipbuilders take overall responsibility for the ship class and subcontract design and production support from parent firms.

---

[23] Obtaining designs from different countries means different design standards. Shipbuilders at the trade level cannot use past experience to resolve issues. Another consideration is that Australia might have to rely on more foreign suppliers with whom it has had minimal prior interaction or direct work experience.

[24] See Thomas Lamb, "Naval Ship Acquisition Strategies for Developing Countries," paper presented at the Pacific Northwest Section Meeting, Society of Naval Architects and Marine Engineers, Vancouver, British Columbia, November 21, 2013. Additional private conversations with experienced past U.S. warship program managers reiterated this point. In addition, Australian shipbuilding practices allow for extended breaks in production, and experienced journeyman leave for higher-paying jobs.

- Connect designers and builders at lead yards by interactive CAD workstations located at the Australian shipyards.
- Evaluate building a first-of-class vessel at an overseas shipyard to leverage the parent shipbuilder's production and design experience, thereby partially mitigating the effects of design errors and changes.

While there may be upfront nonrecurring costs for these approaches, such investments will be more than recovered in efficiencies that are achieved in series production of the ships.

CHAPTER FOUR

Using Indigenous Australian Industry to Address the Short-Term and Longer-Term Gaps in Naval Ship Demand

As described in Chapter Three, there are three potential paths the Commonwealth might take for the indigenous shipbuilding industrial base.[1]

- Path 1 is at one extreme. It would have a fully capable shipbuilding industry that can take an existing or modified design developed elsewhere and build ships to that design in Australia. This is the path used for the *Hobart*-class destroyers.
- Path 2 would be a step down from a full shipbuilding capability, where some portion of a new ship is built in another country and Australian shipyards complete the construction process and install the major weapon and combat systems. This is the path used for the *Canberra*-class LHD program.
- Path 3 would have new RAN ships built and outfitted in another country, with Australian shipyards installing, at most, the final combat system. This is the path that was originally followed for the *Adelaide*-class frigates, the first four of which were built in the United States and delivered to Australia (the last two were built to print in Australian shipyards).

In all three cases, the support of in-service ships would be done in Australia.

---

[1] Of the six paths outlined in Table 3.3, we believe four are viable options. Australia has experience with three of those four; it has not used the combination of building offshore and modifying an existing design but is considering it for the future submarine. In this chapter, we assess only those three options for which Australia has experience.

The analysis of each of these paths depends on the future shipbuilding plans. We define our baseline plan as a drawdown in AWD demand as currently planned and a start of the Future Frigate construction in 2020.[2] From this baseline, we examine the effect of the acquisition plans provided by the two scenarios in Chapter Three and other potential options for lessening the short- and long-term workforce demand gaps. We do this for each of the three potential future paths for the Australian shipbuilding industry.

Analytical Approach, Model, and Assumptions

We adapted RAND's Shipbuilding and Force Structure Analysis Tool to assess the cost and schedule implications of different acquisition and procurement plans.[3] Figure 4.1 shows the basic structure of the tool. We focus on the labor and overhead costs of building the ships under the different procurement plans and therefore use the industrial base model of the overall tool.

The composition of a current fleet, with planned retirement dates, and the desired future force structure define a procurement plan of what types of ships will be built in future years. Each ship has a planned start and delivery date and a demand profile for specific skills during the build cycle.[4]

We use a simulation model to estimate the labor cost and schedule implications of different production plans. Broadly speaking, the simulation models the demand for labor over time and the supply of labor to meet that demand. The model estimates the capacity of the avail-

---

[2] We assume a 2020 start of construction for the first Future Frigate based on a desired in-service date of 2026 to replace the first *Anzac*-class ship and an approximately six-year build and delivery period.

[3] See Mark V. Arena, John F. Schank, and Megan Abbott, *Shipbuilding and Force Structure Analysis Tool: A User's Guide*, Santa Monica, Calif.: RAND Corporation, MR-1743-NAVY, 2004.

[4] The various start and delivery dates and the workload profiles for different classes of ships are provided in Appendix B.

Figure 4.1
High-Level Architecture of RAND's Shipbuilding and Force Structure Analysis Tool

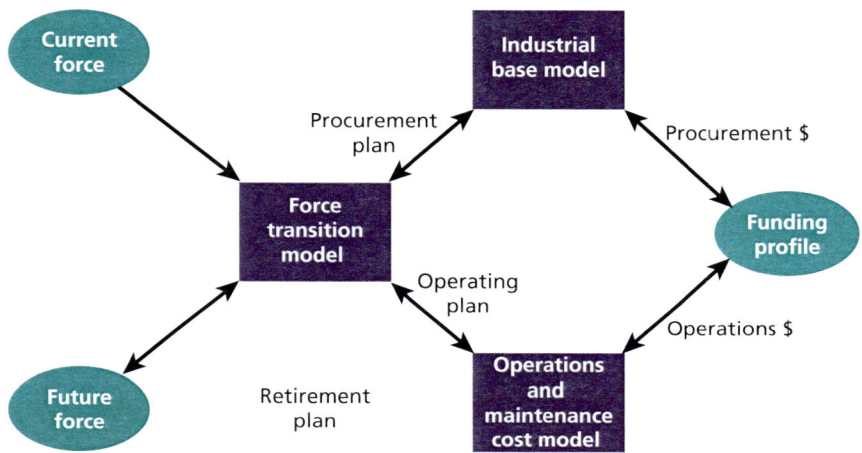

SOURCE: Arena, Schank, and Abbott, 2004.

able labor to meet demand, accounting for such factors as the experience and productivity of the workforce and the capacity to grow the workforce to meet new demands or shrink it to accommodate reduced demands. The model thus provides estimates of the costs and schedule implications of different production plans for different industrial base structures.

The model starts by assuming that the current shipbuilding labor force can meet the current demand. Based on workload input by quarter, the model sees if the existing workforce has the capacity to meet the workload demands (which are assumed to be stated in terms of fully productive workers). If capacity of the current workforce is less than demanded, creating a work backlog, the excess work is pushed to the next quarter and new personnel are hired to meet the next quarter's demands. If existing capacity is greater than the work demanded, excess workers are made redundant. As a result, the model shrinks the workforce when demands are declining and adds to the workforce when demands are increasing.[5] In parallel, the current workforce is aging, gaining proficiency, and losing personnel due to normal attrition. The new employees are drawn from a distribution of proficiency levels and added to the current workforce. Each unskilled worker accounts for unproductive hours until gaining full proficiency. The workforce is broken into five broad skill categories, and calculations are done independently for each skill grouping. The results for each group are combined to provide aggregate results for the workforce.

For analytic convenience, our modeling assumes that the Australian industrial base is constituted by a single "uber" shipyard; we treat the question of how to distribute the shipbuilding work across multiple yards as separate. We first consider a shipyard that builds only large combatants, assuming new patrol boats are built in a different yard; then, we consider a shipyard that builds both patrol boats and combatants.

Two important measures result from the model: the labor cost and the schedule delay of ship delivery that result from different acquisition and production plans. Our labor cost estimates account for varying labor rates across skill categories, a representative model for overhead costs in which the overhead rate decreases with the size of the business base, and the costs of training new hires and terminating redundant workers. Our estimates of schedule assume that work is conducted on a first-in, first-out basis, so ships are not delivered until the shipyards complete all the work that was sent to the yard through the planned delivery date.

Appendix B provides a more detailed description of the assumptions employed in this adaptation of the industrial base model of RAND's Shipbuilding and Force Structure Analysis Tool.

---

[5] Constraints are placed on how many workers a shipyard can shed per quarter and the number of new workers it can hire. These constraints are expressed as percentages of the current workforce.

Addressing the Short Term: Baseline Analysis

AWD block construction and assembly work will end in the next one to three years. As the workforce demand declines, the shipyards, especially those building blocks for the AWD, will begin to shed their workforce. With Future Frigate construction starting in 2020, there is the potential for a gap in demand.

For the baseline analysis, we assume the first-of-class Future Frigate will take 6.5 years to build and will require 5.5 million fully productive man-hours. The second ship in the class will start in 2023, the third ship will start in 2025, and the remaining ships will start one per year after that (approximately matching 30 years from the *Anzac*-class in-service dates). We assume that the second ship will require 5 million fully productive man-hours, and subsequent follow-on ships follow a unit learning curve reduction of 95 percent.[6] Based on these assumptions, Figure 4.2 shows the potential gap in workforce demand, measured by full-time-equivalent (FTE) workers, between the end of the AWD construction program in 2019 and the 2020 start for the first Future Frigate.[7]

The figure highlights how the gap is more significant than the one year implied by the official end of the AWD program and the assumed start of the Future Frigate. Indeed, as many as three to five years may pass between when AWD demand wanes and when production of the Future Frigate ramps up in significant measure. The specific time frame and duration of the gap will vary by skill category.

The peak demand for workers during the production of the Future Frigate is approximately 2,700 skilled personnel. Without some way to lessen the gap between the end of the AWD program and the start of

---

[6] In this report, we define *unit learning curve* as the percentage of man-hours required to construct an additional ship compared with the number of man-hours required to produce the previous ship. Later in this chapter, we will explore the effect of different build schedules and different required levels of effort, and at that time, we will refer to these base assumptions as the *5 million man-hour case*, or *base case*, to simplify our discussion.

[7] This does not include the demand of building patrol boats to replace the *Armidale* class, because in this base case analysis, we assume that the patrol boats are built in a different shipyard than the shipyards building the AWD.

Figure 4.2
Workforce Profile for Building Air Warfare Destroyers and Future Frigates (Base Case)

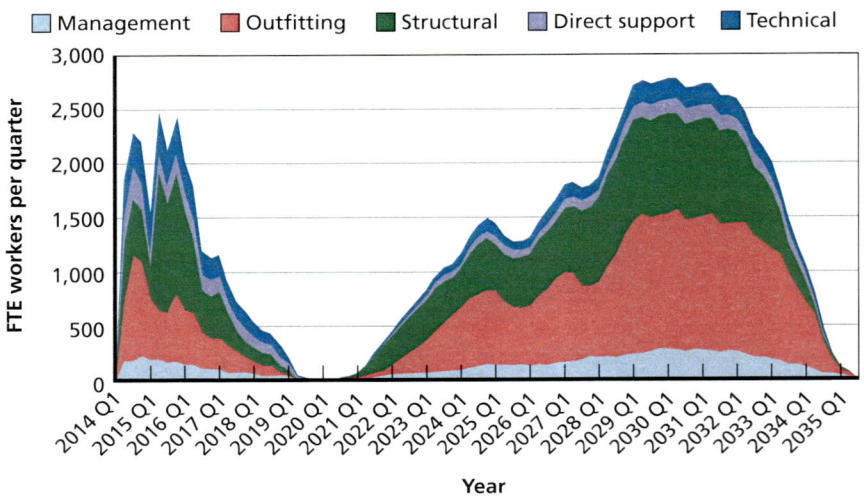

the Future Frigate program, the industrial base will have to build from an almost negligible workforce demand to the 2,700 skilled personnel level in approximately eight years. As a point of comparison, this peak workforce demand is comparable in magnitude to both the build-up for the AWD program and the current workforce levels as shown in Chapter Two, suggesting that there is workforce capacity in Australia to meet the Future Frigate peak demands.

This build-up in workforce will vary by skill category. Table 4.1 shows the peak demands for different skill groups that correspond to the workforce profile in Figure 4.2. Note that these peak demands may occur at different times during Future Frigate production; for example, structural skill demands occur early and decline over time, while the demand for outfitting skills follows a reverse pattern.

Figure 4.3 shows the number of employees that correspond to different percentages of peak demand and how those workforce levels lessen the short-term production gap. Table 4.2 shows an estimation of the percentage of the workforce at each skill level necessary

Table 4.1
Peak Workforce Demands for Future Frigate Construction, by Skill Category (Base Case)

Skill Category	Peak Workforce Demand (FTE workers)
Management	289
Technical	198
Structure	923
Outfitting	1,287
Direct support	151

Figure 4.3
Workforce Profile for Lessening the Short-Term Production Gap (Base Case)

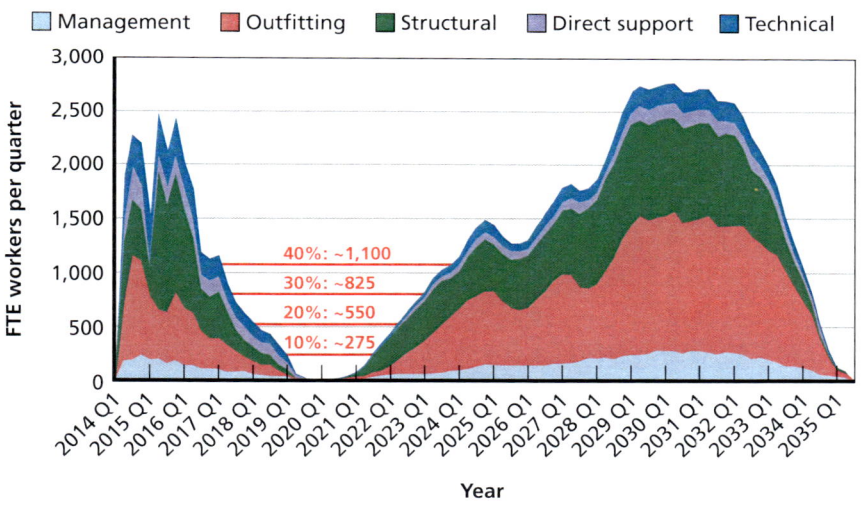

to "balance" the workforce. For example, a workforce of 500 workers should be composed of approximately 50 management, 25 technical, 175 structure, 225 outfitting, and 25 direct support workers.

The primary issue then is to determine what level of construction resources to sustain between the end of the AWD program and the start of the Future Frigate. The Future Frigate program will require a build-up in construction capability—a hill to climb when meeting

Table 4.2
Approximate Skill Percentages of Balanced Workforce Sustainment Levels (Base Case)

Skill Category	Percentage of Total Workforce
Management	10
Technical	5
Structure	35
Outfitting	45
Direct support	5

the desired future production demand. Climbing this hill will require hiring a new, largely unproductive workforce. More workers will be needed to accomplish the same level of workforce demand until the new hires gain full proficiency. The cost of this unproductive labor grows as the percentage of peak workforce demand sustained during the gap is lower. But sustaining higher workforce levels during the gap results in more people to pay. The basic trade-off is between paying some number of workers to do hopefully fully productive and useful work during the gap versus paying for unproductive labor when climbing out of the gap.

Figure 4.4 shows the unproductive man-hours as a function of the percentage of the peak workforce demand sustained during the gap. For example, if 20 percent of the peak demand is sustained, approximately 7 million unproductive man-hours will be included in the Future Frigate build. This is an increase in man-hours of about 20 percent over the approximately 40 million fully productive man-hours to build the ships. As expected, the number of unproductive man-hours drops as larger percentages of the peak demand are sustained.

Figure 4.5 shows the total labor costs by workforce sustainment level of finishing the AWD construction, sustaining various numbers of construction workers during the gap, and then building the Future Frigates.[8] The trade-off shown in the figure is between retaining some level of the workforce, thus providing an experienced base to build

---

[8] The present analysis assumes that the workforce level is sustained in the gap between AWD and the Future Frigate; no workforce level is sustained *after* the Future Frigate.

Figure 4.4
Unproductive Man-Hours, by Workforce Sustainment Level (Base Case)

RAND *RR1093-4.4*

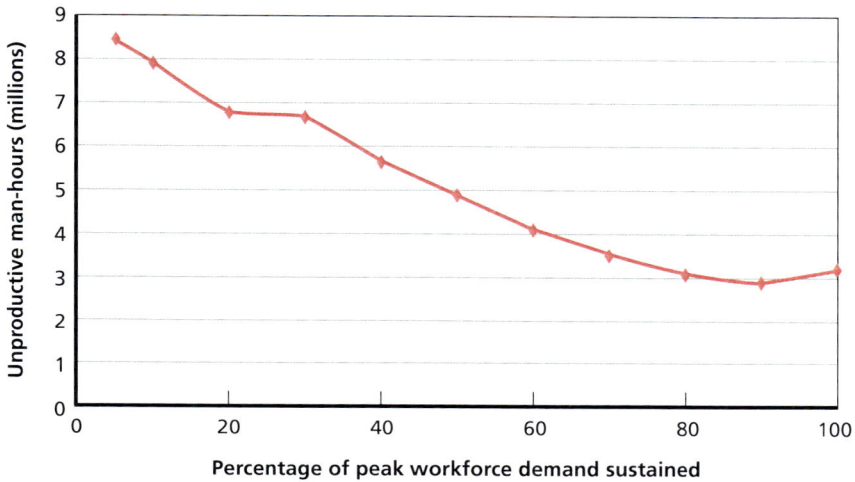

Figure 4.5
Total Labor Costs, by Workforce Sustainment Level (Base Case)

RAND *RR1093-4.5*

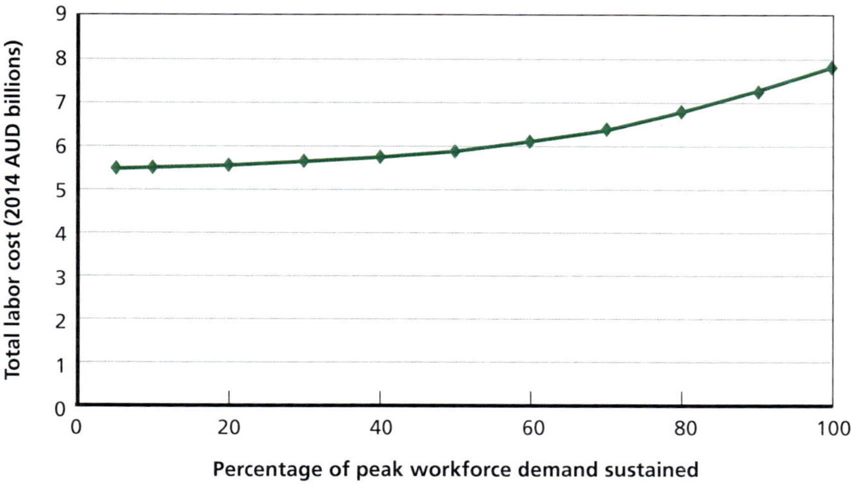

upon, versus making people redundant, thus saving on payroll cost but costing more in building up the workforce to reach proficiency.

One way to interpret Figure 4.5 is that sustaining a workforce of roughly 1,150 productive workers (approximately 40 percent of the peak demand) in the gap between AWD and the planned start of Future Frigate would cost about AUD 300 million more than sustaining roughly 150 workers (about 5 percent of the peak demand), after accounting for the cost of lost productivity in the latter case. The cost of sustaining 575 workers (approximately 20 percent of peak demand) is essentially equal to the cost of sustaining 150 workers (5 percent). If workers sustained in the gap were given productive shipbuilding work, in principle, the workforce would maintain productivity and avoid some of the costs of having to rehire and retrain the workforce that would otherwise be made redundant in a gap. Workforce sustained during the gap would also provide a larger pool of experienced employees to train new workers after the gap, thereby increasing the rate at which the total workforce can grow.

Sustaining a workforce in the gap period has a clear effect on the delivery schedule of the Future Frigates, depending on how many workers are sustained. Figure 4.6 shows these schedule implications if 5 percent of the peak workforce is sustained (roughly 150 workers).[9] The green portions of the bars represent the planned build schedule, and the purple portions are actual delivery dates resulting from any delays in the planned production schedules. The diamonds represent our estimate of planned retirement dates for the *Anzac*-class ships, measured as commissioning date plus 30 operational years (see Table 4.5). When the green or purple bar exceeds the diamond, ships are delivered later than needed to replace a retiring ship. Due to the time to build up the workforce from such a low starting point, the first few ships in the Future Frigate class will take much longer than planned, causing an

---

[9] Our initial model runs assumed no concurrency between the skill sets with regards to schedule (that is, the progress of work in one skill category was not dependent on another). We assume that work is completed on a first-in, first-out basis and that construction of an individual ship is complete when all the work in all skill categories is complete. The impact of this is that schedule delays may be underestimated, and cost could rise due to idling workers waiting for another component to finish, with more schedule delay indicating greater cost.

Figure 4.6
Schedule Implications of Sustaining 5 Percent of Peak Workforce Demand (Base Case)

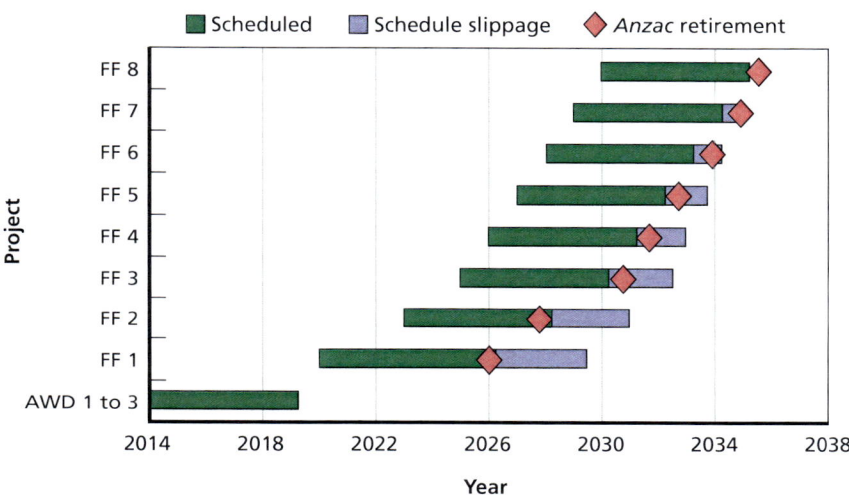

operational availability gap between the retirement of the *Anzac* class and the introduction of the replacement ships. This gap exists almost throughout the build program; that is, almost all Future Frigates (FFs) are delivered later than planned. If the workforce sustainment level is extended to 20 percent of peak demand (roughly 575 workers), as shown in Figure 4.7, there are almost no schedule delays.

The trade-offs between sustaining a certain level of the workforce during the gap and the resulting impact on total labor cost and schedule delays are summarized in Figure 4.8. The red line shows how the total labor cost increases as the workforce percentage increases; the green line shows how the total schedule delay is greatly reduced as the number of workers sustained during the gap increases.[10] If a workforce

---

[10] Total schedule delay is measured in ship-years and corresponds to the total amount of time that *Anzac*-class ship retirements would need to be delayed in order to sustain a constant force structure in spite of delays in delivery of the Future Frigates.

Figure 4.7
Schedule Implications of Sustaining 20 Percent of Peak Workforce Demand (Base Case)

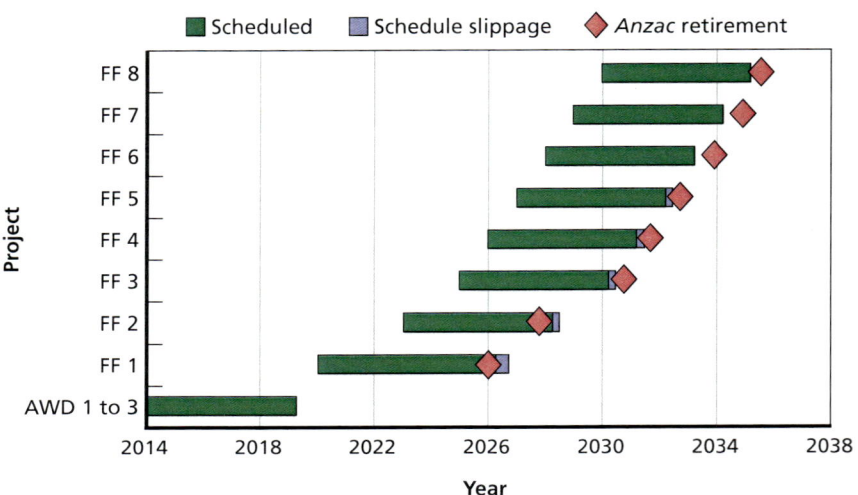

Figure 4.8
Total Labor Costs and Schedule Delays, by Workforce Sustainment Level (Base Case)

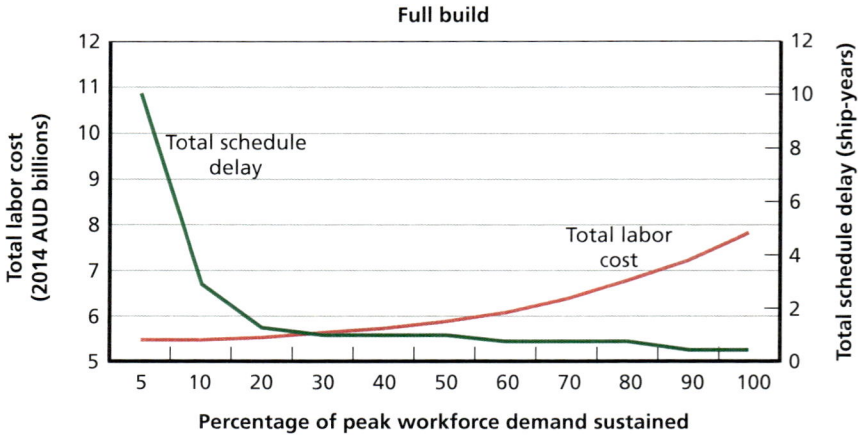

NOTE: Total labor cost is the cost of production for both the new Future Frigates and the remaining AWD program.

between 20 and 40 percent is sustained during the gap, total labor costs rise very little while schedules are largely met.

The foregoing was primarily a theoretical exercise in examining the potential cost and schedule implications of sustaining a workforce in the gap between the end of the AWD program and the start of Future Frigate production. Next, we examine several short-term and long-term options for lessening the gap, and we adopt a continuous build strategy to sustain surface combatant shipbuilding capabilities in Australia.

Sustain a Fully Capable Australian Shipbuilding Industrial Base

Addressing the Short Term

The short-term issues for sustaining a shipbuilding industrial base are the costs and implications for various shipbuilding programs of maintaining workforce sustainment levels between the end of the AWD construction and the start of the Future Frigate build program. We view Australia as having four options for lessening the short-term gap.

Short-Term Options
Start the Future Frigate Earlier
One way to lessen the short-term gap is to shrink its duration. Starting the Future Frigate construction program before 2020 would result in a shorter gap. As mentioned, the Future Frigate program is in the very early stages and has not yet chosen an acquisition path. There are several steps and decisions needed before construction could start. These include deciding whether to modify the common AWD hull or select a modification of an existing overseas option. It will take some time to sort through the advantages and disadvantages of these two options and, if an evolved MOTS hull is chosen, to select the most cost-effective design for the Future Frigate. Once an option is chosen, it will take two or more years to modify the base design to accommodate the new radar and combat management system, support a second helicopter (if the chosen design does not support two helicopters), incorporate new environmental standards, and produce production drawings.

Although starting Future Frigate construction prior to 2020 is highly unlikely, it is informative to show the effect of timely decisions on future programs. We assume a Future Frigate design ready for the start of construction could be optimistically accomplished in three years if decisions are timely and, especially, if a common hull option is chosen. This would suggest an earliest possible construction start of 2018, with delivery as early as 2024. Figure 4.9 shows how the gap is lessened by starting the Future Frigate construction in 2018 instead of 2020.

If the Future Frigate starts in 2018, the workforce sustainment level in the gap increases to roughly 200 workers, or a bit less than 10 percent of the peak demand of the Future Frigate program. This will vary across skill categories, because the types of workers needed at the end of the AWD program and at the beginning of the Future Frigate may be different from those needed during peak construction. Nonetheless, changing the production schedule in this way does not change the amount of work required for the Future Frigates, and thus the preceding workforce sustainment level analysis suggests that costs may decrease a bit by taking advantage of improved workforce pro-

Figure 4.9
Workforce Profile for Building Future Frigates Starting in 2018

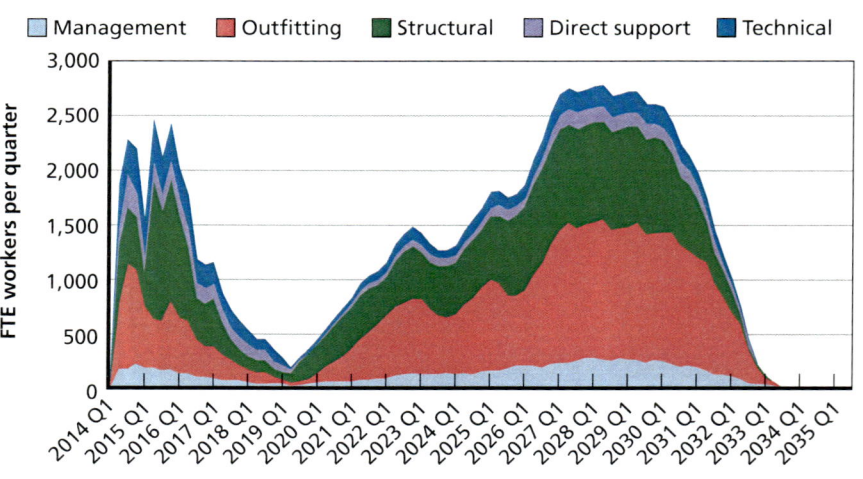

ductivity. If feasible, moving the start of the Future Frigate even earlier will result in more completely closing the gap and in lower overall labor costs.

Figure 4.10 shows the schedule implications if the build of the Future Frigate starts in 2018. Total labor costs are reduced, but the biggest effect is on schedule. Although the schedule slips for a few years relative to planned build durations (reflected by the purple bars), the earlier start of the program allows the Future Frigates to enter service in a timelier manner to replace the *Anzac*-class ships at their planned retirement dates. In fact, the last six ships in the Future Frigate class are delivered before the planned retirements of the *Anzac* ships that they replace. This could allow savings in *Anzac*-class support costs by retiring ships earlier than planned.

Figure 4.11 shows the trade-off between the start of construction for the Future Frigate and the total labor costs and schedule delays in replacing the *Anzac*-class ships. Although starting construction of the Future Frigate one or two years prior to the assumed 2020 start will reduce both labor costs and schedule delays, the knee in the curve at

Figure 4.10
Schedule Implications of Building Future Frigates Starting in 2018

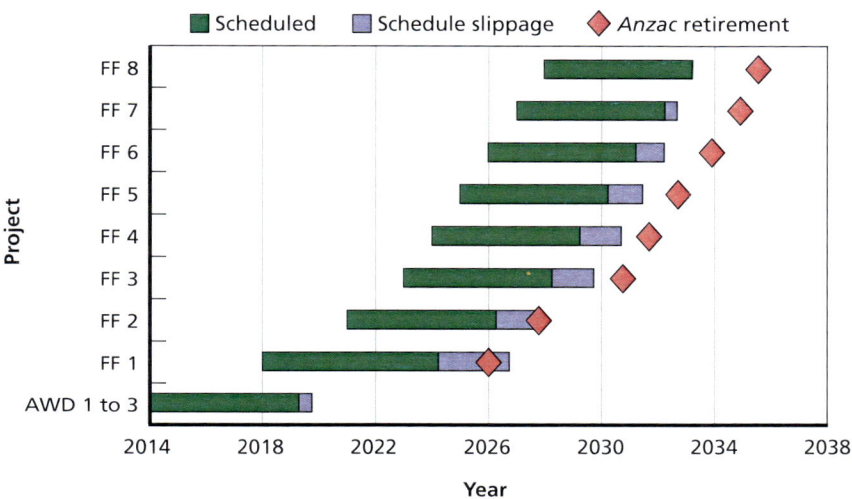

RAND RR1093-4.10

Figure 4.11
Total Labor Costs and Schedule Delays, by Future Frigate Construction Start Date (Full Capability Path)

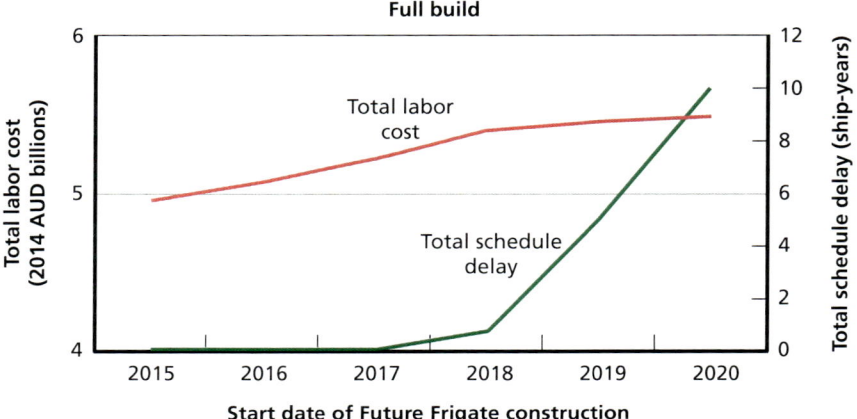

NOTE: Total labor cost is the cost of production for both the new Future Frigates and the remaining AWD program.
RAND RR1093-4.11

2018 suggests even greater cost savings would result from starts before 2018. Although starting construction before 2020 is highly unlikely, the figure shows the need for timely decisions that look well into the future.

Nonetheless, while starting construction before 2020 can reduce costs and schedule delays, starting earlier means ending the build program earlier. Completing the Future Frigate program, coupled with the likely start of the next surface combatant program (to replace the AWDs), would create a gap similar to the one facing Australian shipbuilders today. Sustaining a fully capable shipbuilding industrial base will require a continuous build strategy that avoids gaps in demand. We will address such a continuous build strategy later in this chapter.

Build a Fourth Air Warfare Destroyer

A second option for lessening the gap between the end of the AWD build and the start of the Future Frigate is to build a fourth AWD. With a timely award of the fourth ship, the shipbuilders can transition from finishing the third ship to beginning the fourth, and with proper planning, they can sustain their workforce until the Future Frigate program begins. The question is whether this fourth AWD is needed (including the ability to recruit a crew for the ship) and whether the much higher cost of acquiring the fourth AWD outweighs the potential money saved by sustaining the workforce. We assume that building the fourth AWD reduces the Future Frigate buy to seven ships and that it would replace the first *Anzac* ship planned to retire in 2026.

Figure 4.12 shows how adding a fourth AWD lessens the gap between the first three AWDs and the 2020 start of the Future Frigate build. Figure 4.13 shows the effect on the Future Frigate schedule. Adding the fourth AWD increases workforce efficiency by sustaining the workforce in the gap, but total labor cost increases slightly because

Figure 4.12
Workforce Profile for Adding a Fourth Air Warfare Destroyer

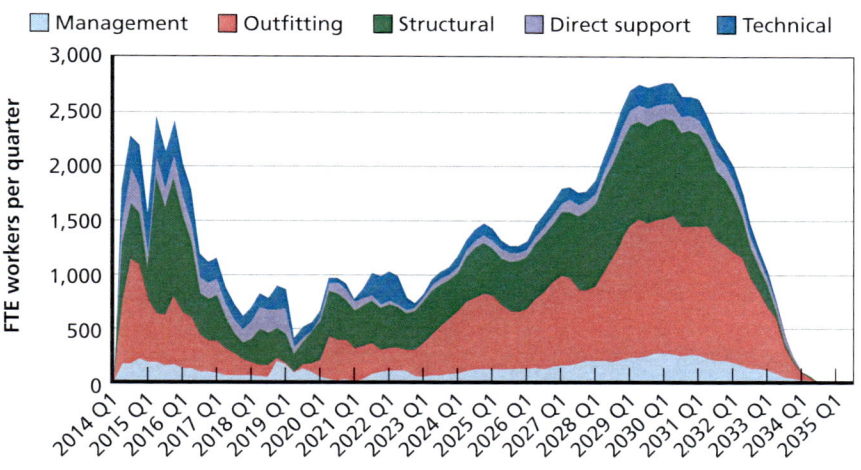

Figure 4.13
Schedule Implications of Adding a Fourth Air Warfare Destroyer

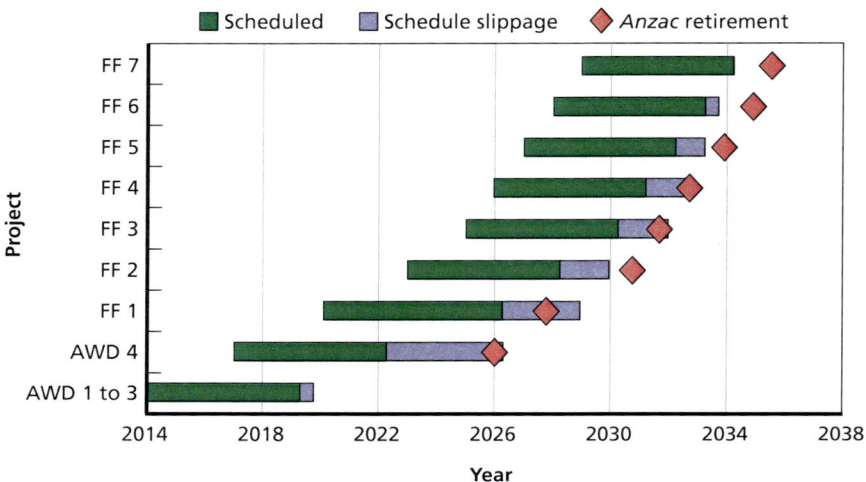

of the larger ship.[11] Although there is schedule slippage with this option relative to planned build durations, the fourth AWD enters service in time to replace the first *Anzac* ship, and the Future Frigate deliveries match closely with *Anzac*-class retirements.

There is no stated requirement for a fourth AWD. Although there are modest labor cost savings, labor represents less than half of the total cost of a major surface combatant. Given the lack of a requirement and the high cost of the ship, it is unlikely that building a fourth AWD is a feasible option.

Build Patrol Boats in the Major Shipyards

A third option for lessening the workforce demand gap is to build smaller boats in the major shipyards during the gap. As discussed, there

---

[11] As detailed in Appendix B, our baseline analysis assumes that the fourth AWD will require 5.5 million fully productive man-hours and 22 quarters to build. Even before accounting for the effects of unit learning curve, this is more than 500,000 man-hours greater than what we assume will be required for the eighth Future Frigate that would otherwise be produced without the introduction of a fourth AWD.

are two acquisition scenarios under consideration (see Table 3.2). In Scenario 1, the shipyards would build 14 patrol boats, with in-service dates starting in 2021; in Scenario 2, the shipyards would build six to eight patrol boats and switch to building OPVs/LMRVs sooner.

In our previous analyses, we assumed that patrol boats would be built in Australia's small shipbuilding yards. Here, to lessen the gap, we assume that the shipyards that build large combatants would build the patrol boats between the end of the AWD program and the start of the Future Frigate build. Figure 4.14 shows the effect on workforce numbers of building 14 patrol boats with in-service dates between 2021 and 2026 (Scenario 1). Given the short time required to build the patrol boats, we start construction in 2020 (for delivery in 2021) and build two boats in 2020, 2021, 2024, and 2025 and three patrol boats in 2022 and 2023 (for a total of 14 boats).[12] We assume that

Figure 4.14
Workforce Profile for Building Patrol Boats at Major Shipyards Starting in 2020

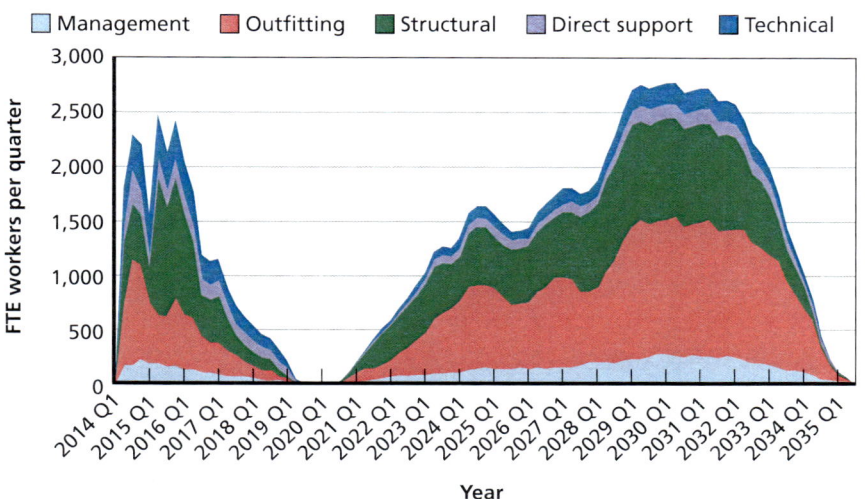

---

[12] As described in Appendix B, we assume that production of the patrol boats will require 140,000 fully productive man-hours and five quarters.

construction on the Future Frigates starts in 2020. The net effect of a 2020 start is to push the patrol boat construction on top of the Future Frigate demand, thus increasing the peak workforce demand without increasing the demand in the gap period. Because of this, we did not analyze this option any further. However, starting the patrol boat construction in 2017 will have an effect on lessening the workforce demand gap (see Figure 4.15). This early start of patrol boat construction assumes that a suitable design is available and that the contracting process experiences no delays.

Figure 4.16 shows the schedule implications of adding patrol boat construction to the major shipyards starting in 2017. Because adding the patrol boats sustains only slightly more than 5 percent of the workforce, the first three Future Frigates are delivered much after the planned retirement dates of the *Anzac*-class ships that they are intended to replace. Figure 4.16 also shows that producing patrol boats in the same shipyard as large combatants would expose the patrol boats

Figure 4.15
Workforce Profile for Building Patrol Boats at Major Shipyards Starting in 2017

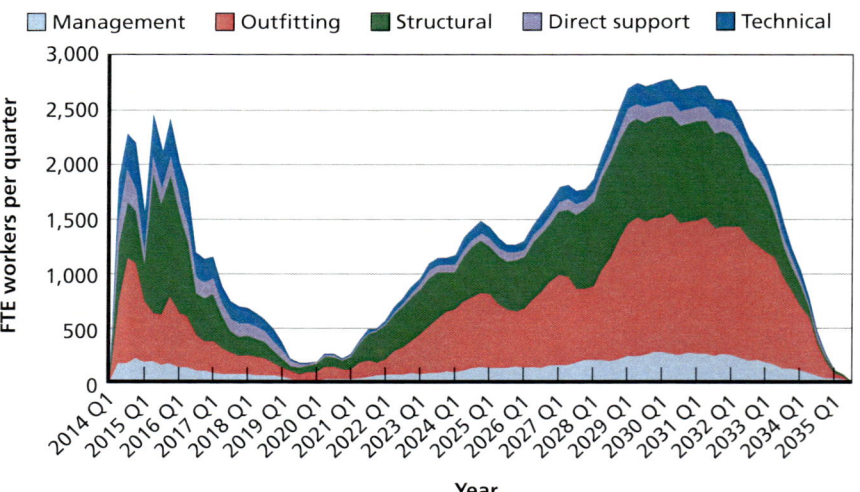

Figure 4.16
Schedule Implications of Building Patrol Boats at Major Shipyards Starting in 2017

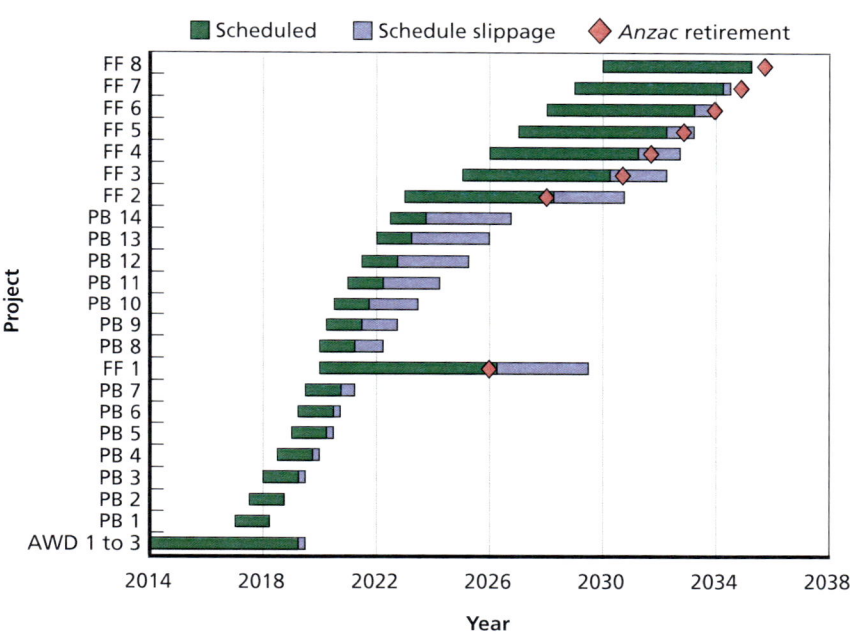

to a risk of delays, because both ship types will compete for a common workforce and facilities.

In Scenario 2, the number of patrol boats built in major shipyards would be reduced to between six and eight, and an LMRV program of 21 ships would begin, with in-service dates starting in 2026. Because the Future Frigate will start around 2020 and will continue for approximately 12 years or more, this scenario adds demand on top of the Future Frigate program while doing nothing to help sustain the workforce in the short-term gap. Furthermore, there is no official program or stated requirements for this new class of ships. For these reasons, we do not evaluate an early build of a littoral multirole class as an option for lessening the short-term gap.

Build Offshore Patrol Vessels in the Major Shipyards

All of the previous options for lessening the short-term gap have various disadvantages.[13] A potentially more viable option is to build a number of OPVs starting in the next two years. The United Kingdom is taking this approach by building three OPVs to lessen the gap between the end of the *Queen Elizabeth* aircraft carrier program and the start of construction of the Global Combat Ship (the Type 26). There are several existing OPV designs available, including the one being built in the United Kingdom. Assuming that an existing design is chosen with little or no modifications, the start of OPV construction could begin in 2017.

For purposes of this analysis, we assume a workforce demand profile of 0.7 million man-hours (approximately five times the demand profile we used for the patrol boats) over 12 quarters of construction.[14] Figure 4.17 shows how building four OPVs would lessen the short-term

Figure 4.17
Workforce Profile for Building Four Offshore Patrol Vessels at Major Shipyards

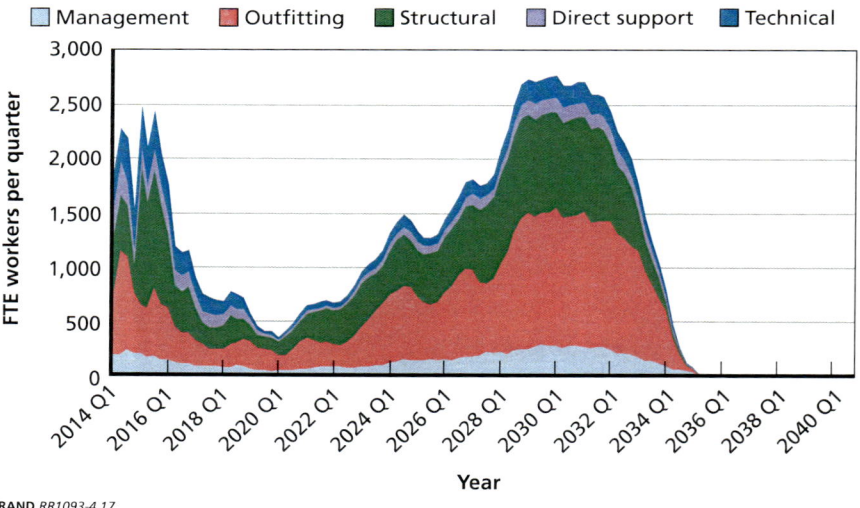

---

[13] A sensitivity analysis that varies several of our baseline OPV assumptions is presented in Appendix D.

[14] This ship profile is derived assuming an OPV with displacement of 1,700–1,800 metric tons and length of 80–90 m.

gap. We assume construction starts in 2017, with a basic drumbeat of approximately one. (From a shipbuilding perspective, the drumbeat refers to how frequently new ships are delivered to RAN.)

Figure 4.18 shows the total labor costs of finishing the AWD construction, building the eight Future Frigates, and building various numbers of OPVs in the gap. The red bar represents the labor costs (AUD 5.49 billion) of the base case—starting the build of the Future Frigates in 2020, with a drumbeat of one for the last six ships. The other bars are the labor costs of building three, four, or five OPVs in the gap, with the blue portion representing the base case costs and the purple portion representing the additional labor costs of the OPVs. For example, building four OPVs adds AUD 130 million to the base case labor costs. In essence, the four OPVs could be built basically for "free," given that they are sustaining productive labor that reduces the costs of unproductive labor when building the workforce for the Future Frigate construction.

Figure 4.18
Total Labor Costs of Building Three, Four, or Five Offshore Patrol Vessels at Major Shipyards

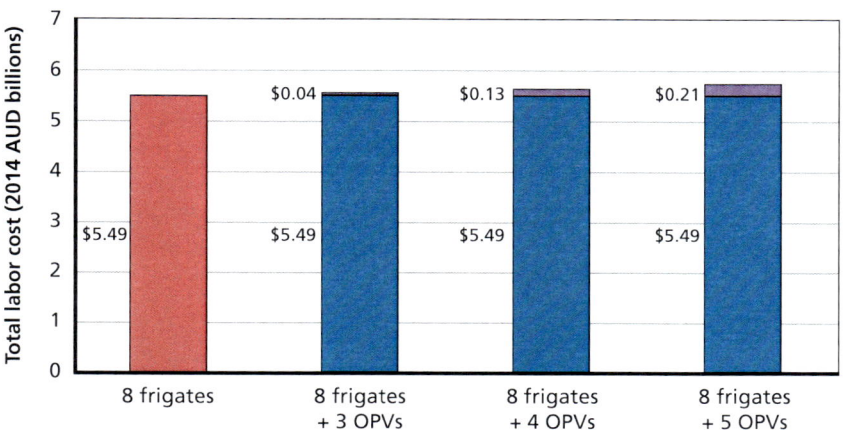

NOTES: Red and blue connote baseline labor costs for eight Future Frigates, assuming a one-year drumbeat. Purple connotes the marginal OPV-specific labor costs involved in building different numbers of OPVs in addition to building Future Frigates on that one-year drumbeat.

Figure 4.19 shows the effect on schedule delay in delivering Future Frigates to replace the *Anzac* class when adding different numbers of OPVs in the gap. The base case of starting the Future Frigate build in 2020, with a one-year drumbeat, results in a delay of ten ship-years. Lessening the gap with OPVs reduces that delay to less than two ship-years.

Sensitivity Analysis of Future Frigate Workforce Demands

The cost of sustaining a workforce during the gap and then using that workforce to serve as the foundation for building the Future Frigates depends on how high a hill to climb—that is, the workforce demand profile for the Future Frigate program, the start date of construction, and the duration of the build. The analysis in the four options we just evaluated assumes that the first Future Frigate would start construction in 2020 and require 5.5 million man-hours of productive labor over a

Figure 4.19
Total Schedule Delay Relative to *Anzac*-Class Retirements of Building Three, Four, or Five Offshore Patrol Vessels at Major Shipyards

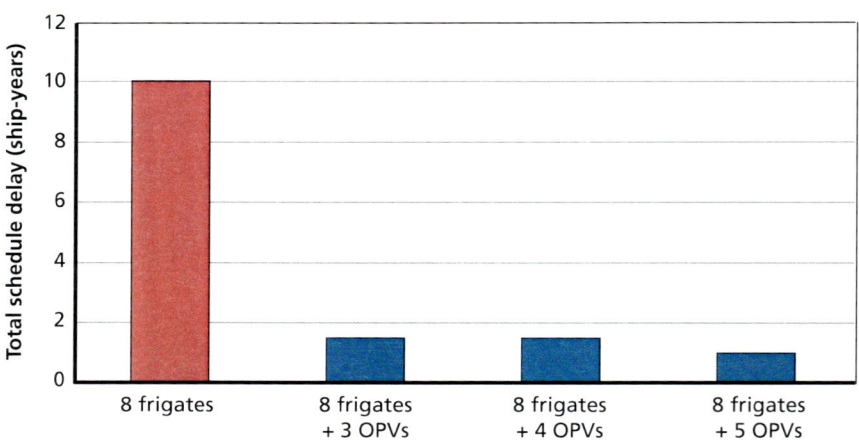

NOTE: Schedule delay is for delivering eight Future Frigates, assuming a one-year drumbeat.
RAND RR1093-4.19

6.5-year build period.[15] In Appendix C, we examine the effect of other demand profiles that vary the level of effort to produce a frigate, the drumbeat at which frigates are produced, and the unit learning curve, among other variables. In the paragraphs below, we examine the broad implications of this sensitivity analysis.

There is typically a relationship between the total workforce demand and the build duration—more work normally requires more time (or a more rapid growth in workforce). If the total required workforce demand for the Future Frigate is reduced and that work is accomplished in a shorter period of time, the peak demand function may not change significantly. The same is true if the workload for the Future Frigate and the time to build the ship are increased. However, workforce demands greater than our base case assumption of 5.5 million man-hours will lead to higher total labor costs, and lower demands will lead to lower total labor costs.

Several broad observations can be made from this sensitivity analysis. In general, the demand variables follow predictable trends. Larger levels of effort will be more costly and risk additional delays; longer drumbeats can increase delays and may save money by reducing peak demand if they are not too long (in this analysis, a 1.5-year drumbeat struck a balance between one- and two-year drumbeats, although the differences were small in an absolute sense); and unit learning rate has no significant effect, given the relatively small quantity of ships produced.

What's more interesting is how the relative attractiveness of options remains largely unchanged by these variables. First, regardless of level of effort, drumbeat, or unit learning curve, the best option from both a cost and schedule perspective is starting production of the frigates early. In general, starting production in 2018 saves money and increases the chance of delivering ships in time to replace the retiring *Anzac* class; the schedule impact is largely a result of having two addi-

---

[15] As discussed, our 5 million man-hour base case analysis assumes that 5.5 million fully productive man-hours are required for the first-of-class Future Frigate, and then 5 million man-hours and a 95-percent unit learning curve are required for the second ship and subsequent follow-ons.

tional years before the first retirement. There can be a notable difference even between starting production in 2017 versus 2018, although as noted elsewhere in this report, it is likely impractical to start production before 2020, given the considerable design and contracting work that remains to be done for the Future Frigate program. If production cannot be started until 2020 or later, it appears that steps will be needed to mitigate cost and schedule implications of a production gap.

Second, in most cases, the option of adding a fourth AWD increases the overall productivity of the workforce and mitigates delays in delivery, but it increases total labor costs. Our analysis suggests adding a fourth AWD *could* be cost-competitive from a labor perspective if the Future Frigate is the largest of the variants explored here (7 million man-hours), but even in this case, the cost savings would only apply if the fourth AWD replaces one of the eight frigates. As a general conclusion from examining the many cases presented here, adding a fourth AWD can mitigate risks to schedule, although the specific effects will depend on the level of effort required by the Future Frigate.

Third, building the patrol boats in the major shipyards can improve productivity, and if the patrol boat construction starts in 2017, there are very modest savings compared with the base case.[16] However, this option does not fully mitigate the effect of delayed delivery without also starting production of the Future Frigates by 2018. Steps may be needed to ensure that the patrol boats themselves are not delayed after the start of the frigates.

Finally, producing three to five OPVs appears to be an effective way to mitigate the effects of a workforce demand gap. The labor cost of producing these additional ships is largely offset by the savings that stem from sustaining a productive workforce. Adding OPVs also mitigates production delays that arise from leaving the gap unfilled.

A range of other factors—such as hiring rate, workforce ceiling, and the rate at which new workers gain proficiency—could well affect cost and schedule outcomes. However, variables such as these are not

---

[16] These savings come from building the patrol boats in the same shipyards where the Future Frigates will be built versus in other Australian shipyards. Money will be spent to build the ships either way, but cost varies depending on shipyard.

Summary of the Implications of Short-Term Options

Analysis of the base case suggests that there is a short-term gap in workforce demand between when the AWD program ends and the Future Frigate program begins. Sustaining some portion of the productive workforce during that gap will be needed for the start of the Future Frigate construction. The costs of rebuilding the workforce to meet future demands increases as the workforce sustainment level decreases. More unskilled labor must be hired and trained, resulting in an increase in nonproductive man-hours. But sustaining a large workforce base implies paying their salaries and finding productive work for them. The biggest impact is on the delivery schedule of the Future Frigates, as smaller sustained workforces lead to longer construction periods.

There are few options available to cost-effectively sustain the shipbuilding production workforces before the start of the Future Frigate program. We have examined some of these options in the analyses in this chapter. Table 4.3 summarizes various cost and schedule measures for the strategies we examined, as well as for other strategies. The table shows the total labor costs and total schedule delay (relative to planned *Anzac*-class retirements) of completing the AWD build program, building the Future Frigates, and building the various options examined for lessening the gap. For example, when sustaining a workforce equivalent to 5 percent of the peak demand during the Future Frigate program, we estimate the total labor cost of the base case at AUD 5.49 billion (which includes AUD 120 million for the patrol boats built in another shipyard).[17] The total delay (that is, cumulative across all eight ships in the program) relative to planned *Anzac* retirements is ten years.

---

[17] The cost estimates shown in Table 4.3 and other tables are not budget quality. It is better to consider the relative costs between options rather than the absolute costs.

Table 4.3
Summary Labor Costs of Various Options for Workforce Sustainment (Full Capability Path)

Option	Total Labor Costs (2014 AUD billions)	Total Schedule Delay Relative to *Anzac*-Class Retirements (years)
Base case: 3 AWD, 8 FF (2020)	5.49[a,b]	10.00
3 AWD, 8 FF (2018)	5.40[a]	0.75
4 AWD, 7 FF (2020)	5.54[a]	1.25
4 AWD, 7 FF (2018)	5.39[a]	2.25
3 AWD, 8 FF (2020), 14 PBs (2017)	5.50	8.50
3 AWD, 8 FF (2018), 14 PBs (2017)	5.34	0.00
3 AWD, 8 FF (2020), 3 OPVs (2017)	5.52[a]	1.50
3 AWD, 8 FF (2020), 4 OPVs (2017)	5.61[a]	1.50

[a] Includes the cost of building 14 patrol boats starting in 2017 at different shipyards. We estimate the labor costs of the 14 patrol boats at AUD 120 million.

[b] Assumes a 5-percent workforce sustainment level in the gap period.

Direct labor cost represents the wages paid to the workers each quarter summed across all quarters, accounting for different labor rates across skill categories. Overhead costs reflect the shipyard overhead adjusted on the basis of number of workers employed (overhead rates go up as workforce demand goes down.) Training and termination costs are associated with training new employees until they are fully productive and terminating employees when there are excess workers (basically severance costs).

The table shows that total labor costs do not change much (from AUD 150 million cheaper than the base case to AUD 120 million more expensive) across the various options but that most options for lessening the gap have a significant effect on total delay in delivering *Anzac*-class replacements. Also, lessening the gap with OPVs provides additional ships to RAN at a very marginal labor cost to produce them.

Addressing the Longer Term

When considering sustaining a fully capable Australian shipbuilding industrial base, the longer-term question concerns the construction plans after the Future Frigates are all built. The first acquisition scenario defines the need for 21 LMRVs with in-service dates starting in 2035 (see Table 3.2). We assume that 500,000 man-hours over eight quarters are needed to build the ships, construction starts in 2033, and two ships are ordered every year until 2043. Figure 4.20 shows our estimate of the future workforce demand of the first scenario. The figure shows that there is no gap between the end of the Future Frigate program and the start of the LMRV build, although a small gap would exist if the Future Frigate program began construction in 2018 rather than 2020. That gap could be closed by starting the LMRV earlier. The larger concern is the difference in heights of the Future Frigate and LMRV demand curves. The LMRV will require a maximum of approximately 500 shipbuilding workers, compared with almost 3,000 for the build of a major surface combatant. As discussed with the short-term gap, there are various costs to consider when building a workforce

Figure 4.20
Workforce Profile for Scenario 1, Longer-Term (Full Capability Path)

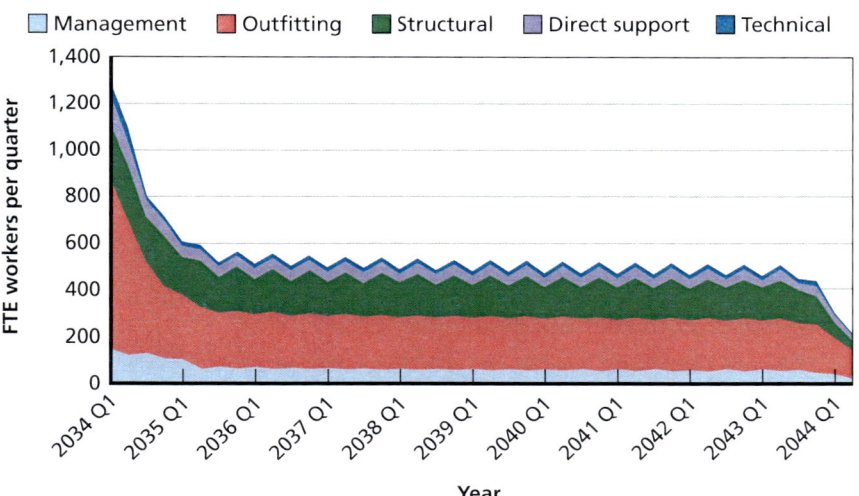

to meet a new peak demand. Sustaining a base of approximately 500 skilled workers implies about 15 percent of the future peak demand.

The 21 littoral multirole ships will provide some demand for new ship construction after the Future Frigates. However, they are not large, complex surface combatants. Given the new AWDs, LHDs, and Future Frigates and their expected 30-year operational lives, there may not be the start of a new major surface combatant construction program from the end of Future Frigate construction in 2034 until a new build program in the mid-2040s. A continuous build strategy would seek to level load demands over this time period and beyond. In the previous analyses, we assumed that the delivery of the Future Frigates was timed to match the retirement of the *Anzac*-class ships. A second option is to spread out the build of the Future Frigates to correspond to a drumbeat that would provide a continuous build plan to sustain the Australian shipbuilding industrial base in a cost-effective manner.

From a shipbuilding perspective, the drumbeat refers to how frequently new ships are delivered to the Navy. For example, a drumbeat of one implies that a new ship is delivered each year. In the short term (2015 to 2030), the drumbeats are determined by the need to replace ships currently in the RAN force structure. For example, the last six *Anzac*-class frigates were commissioned at the rate of one per year, suggesting the new frigates that will replace the *Anzac* class will be needed at the same rate (i.e., a drumbeat of one).

There is a direct relationship between the size of the future force, the drumbeat at which new ships enter service, and the average operational life of the ships. In general, the drumbeat equals the size of the force divided by the average ship life. If ships last 30 years on average and the desired force structure is 30 ships, then a new ship is needed every year. The force structure resulting from different drumbeats and average operational ship lives is shown in Table 4.4.

The currently planned naval force structure includes three AWDs, eight to ten Future Frigates, two LHDs, and an LSD, for a total of 14 to 16 major surface ships. There are also plans for 27 to 35 smaller patrol boats, OPVs, and LMRVs. Drumbeats greater than two (i.e., delivering a new ship at a slower pace than every two years) will probably not sustain the desired future force structures.

Table 4.4
Force Structures for Different Drumbeats and Ship Lives (Full Capability Path)

Drumbeat (years)	Months Between Construction Starts	Force Structure (number of ships)				
		20-Year Ship Life	25-Year Ship Life	30-Year Ship Life	35-Year Ship Life	40-Year Ship Life
1	12	20	25	30	35	40
1.5	18	13	17	20	23	27
2	24	10	13	15	18	20
2.5	30	8	10	12	14	16
3	36	7	8	10	12	13

The green cells in Table 4.4 highlight the drumbeat and operational life combinations that meet Australia's currently planned large ship naval force structure. Two such combinations would work for average ship lives of 25 and 30 years. Two other combinations would work for operational lives of 35 or 40 years. However, it is unlikely that RAN will operate ships for more than 30 years. Therefore, based on the desired future force structure, a drumbeat of 1.5 or 2.0 would seem appropriate.

The shipbuilding drumbeat and the duration and man-hours to build a ship will determine the annual shipbuilding workforce demand for the industrial base. The man-hours to build a ship normally build up slowly in the beginning of the build period, level off for a short period of time, and then drop as the ship nears completion. As more ships in the class are built, individual ship demands overlap to present a longer-term construction demand. The desire is for a workforce profile that minimizes the peaks and valleys in demand. For larger drumbeats, the individual ship demand functions begin to separate, creating higher peaks and deeper valleys.

Figure 4.21 shows the workforce demand profile for starting the build of the first Future Frigate in 2020, the second ship in 2023, and the remaining six ships in the class following a drumbeat of two.[18] With an assumed 30-year operational life for the AWD, the first AWD

---

[18] Also included is the LMRV demands from Scenario 1, with construction starting in 2033.

Figure 4.21
Workforce Profile for a Continuous Build of the Last Six Future Frigates with a Drumbeat of Two

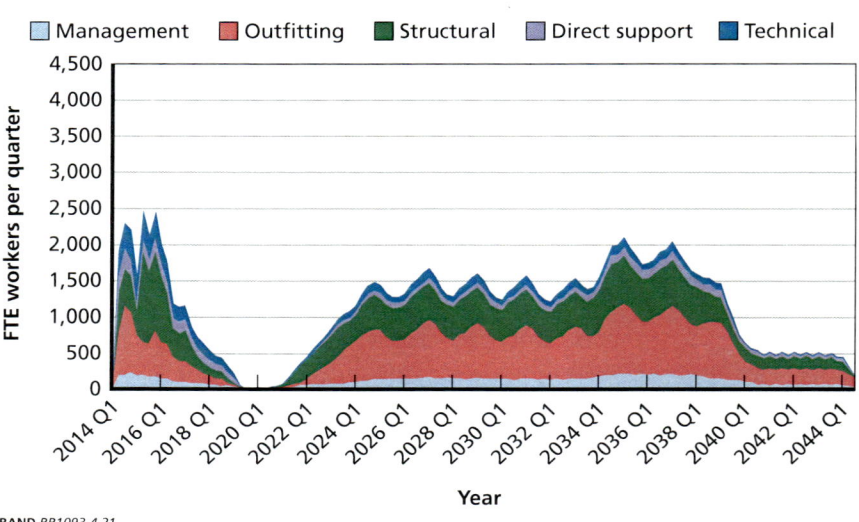

will leave service in approximately 2047. Assuming a six-year build for the replacement ship, construction would start in 2041, at approximately the same time as the build of the Future Frigates would end (assuming the drumbeat of two). A drumbeat of 1.5 would end Future Frigate construction a few years before the start of the AWD replacement, thus creating a short gap in workforce demand.

Although building the Future Frigates at a drumbeat of two versus a drumbeat of one will provide a continuous build strategy that sustains an Australian shipbuilding industrial base, it has an effect on the transition from the *Anzac* class to the new class. Table 4.5 shows the transition from the *Anzacs* to the Future Frigates with a drumbeat of two. The table shows the commissioning and retirement years for the eight *Anzac* ships and the start of construction and commissioning year for the Future Frigates. It also shows the gap between when an *Anzac* ship is scheduled to retire and when the replacement is commissioned

Table 4.5
Future Frigate Force Structure with a Drumbeat of Two, 2026–2041

Ship	Commission Year	Retirement Year[a]	Start Replacement Ship Construction	Replacement Ship Commissioned[b]	Operational Availability Gap
Anzac	1996	2026	2020	2026	0
Arunta	1998	2028	2023	2029	1
Warramunga	2001	2031	2025	2031	0
Stuart	2002	2032	2027	2033	1
Parramatta	2003	2033	2029	2035	2
Ballarat	2004	2034	2031	2037	3
Toowoomba	2005	2035	2033	2039	4
Perth	2006	2036	2035	2041	5

[a] Assuming 30 years from commissioning.

[b] Assuming six years to build and commission.

(assuming a six-year build period).[19] Because the past six *Anzac* ships were delivered with a drumbeat of one, a drumbeat of two causes the operational availability gap between retirement and replacement to grow for the last six *Anzac* ships. These yearly gaps could be closed to some degree by extending the life of the *Anzac* ships or by reducing the time to build and commission the Future Frigates (assumed to be six years).

Figure 4.22 shows how the RAN frigate force structure changes if the Future Frigates are built with a drumbeat of two. From 2026, when the first *Anzac* retires, there is basically no change in the force structure until 2031, when the third *Anzac* ship retires. However, the future RAN frigate force structure will drop from eight to five from 2031 to 2036 before starting to build back up in 2037. The force structure will not return to eight frigates until 2041 when the last Future Frigate is delivered.

A continuous build strategy for a fully capable Australian shipbuilding industrial base must consider both short-term and longer-

---

[19] The gap between delivery and retirement of an *Anzac* ship will be greater due to the schedule slippages indicated previously. This difference will depend on the option chosen for lessening the workforce demand gap and will be greatest for the earlier ships in the class.

Figure 4.22
Future Frigate Force Structure with Drumbeat of Two, 2026–2041

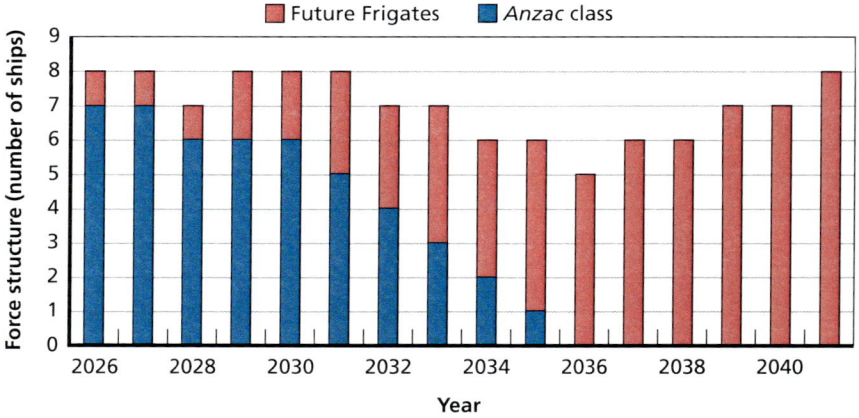

term options. Figure 4.23 shows the total labor costs when building various numbers of OPVs in the short-term gap (one of the more promising short-term options) and adopting a drumbeat of two for the Future Frigates. The red bar represents the base case total labor cost (AUD 5.49 billion) of starting construction of the Future Frigate in 2020 with a drumbeat to match *Anzac*-class retirements. The other bars represent the total labor costs of building the eight Future Frigates with a drumbeat of two and various numbers of OPVs built during the short-term gap. For example, moving to a drumbeat of two adds approximately AUD 30 million to total labor costs. If four OPVs are built in the short-term gap, labor costs add only AUD 190 million to that increase while providing the labor to build the OPVs.

The real impact of a Future Frigate drumbeat of two is on the size and composition of the RAN Future Frigate fleet. Figure 4.24 shows the total schedule delay in ship-years of the base case and the various options for building Future Frigates and OPVs. As discussed previously, the base case of starting the build of the Future Frigates in 2020 with a planned delivery schedule that matches *Anzac*-class retirements results in a delay of ten total ship-years because of the low productivity of the workforce. Moving to a drumbeat of two for the Future Frigates

Figure 4.23
Total Labor Costs for Building Three, Four, or Five Offshore Patrol Vessels in the Short Term, with a Future Frigate Drumbeat of Two

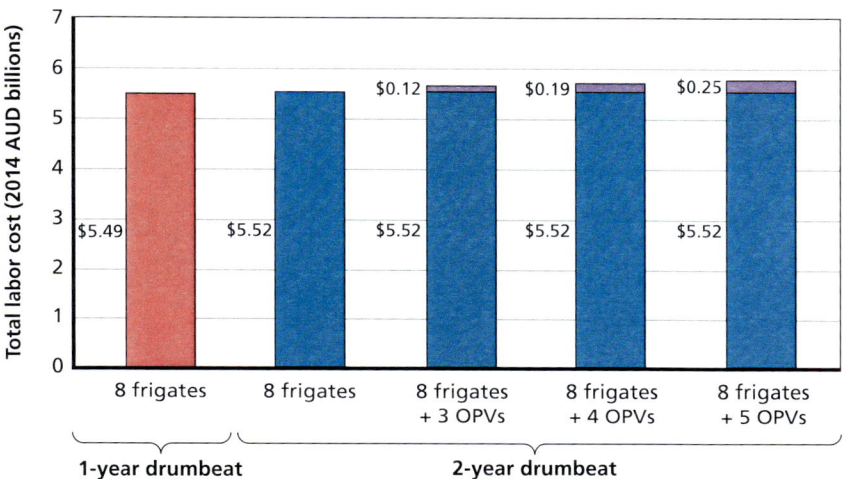

NOTES: Red connotes the baseline labor cost for eight Future Frigates, assuming a one-year drumbeat. Blue connotes Future Frigate labor cost assuming a two-year drumbeat. Purple connotes the marginal OPV-specific labor costs involved in building different numbers of OPVs in addition to building Future Frigates on that two-year drumbeat. The figure assumes that Future Frigate construction starts in 2020, with a two-year drumbeat after the second hull, and that the unit learning curve is 95 percent.
RAND RR1093-4.23

increases that delay to approximately 24 years and results in a Future Frigate force that is one to three ships less than the desired force structure of eight for several years (see Figure 4.22). Adding OPVs in the short-term gap can reduce some of the impacts of the increased drumbeat but still results in a total delay of approximately 14 ship-years.

In the long term, a continuous build strategy of building major surface combatants with a drumbeat of two should sustain a healthy and cost-effective shipbuilding industrial base. Building OPVs during the short-term gap will provide a cost-effective transition to the lower workforce demands of a Future Frigate program using a drumbeat of two. And the end of the Future Frigate build program would flow into the build of the next major surface combatant. There will be challenges

Figure 4.24
Total Schedule Delay for Building Three, Four, or Five Offshore Patrol Vessels in the Short Term, with a Future Frigate Drumbeat of Two

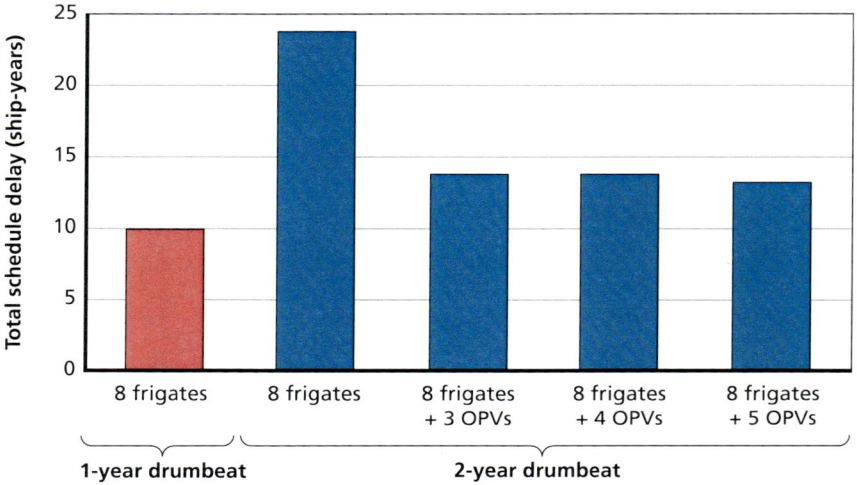

NOTE: The figure assumes that Future Frigate construction starts in 2020, with a two-year drumbeat after the second hull, and that the unit learning curve is 95 percent.
RAND RR1093-4.24

during the replacement of the *Anzac* class, but these challenges might be overcome with careful planning of delivery schedules and usage of the existing fleet.

Structure of a Fully Capable Australian Shipbuilding Industrial Base

The analyses of various options to sustain the capability to build new ships in Australia did not make a distinction between different shipyards but rather focused on the total workforce demand. The AWD is a shared build, where three shipyards build portions of each ship and one shipyard integrates the various pieces and delivers the ships to RAN.[20]

---

[20] Shared build strategies have been used in other programs for various reasons. Examples include the U.S. *Virginia* and *Zumwalt* programs, the U.K. Type 45 and *Queen Elizabeth* programs, and the French *Mistral* program. Smallman et al. (2011) provide details on these and other examples of shared build programs and describe the various advantages, disadvantages, and costs of such strategies.

For future shipbuilding programs, decisions will be needed on how many shipyards to sustain in Australia and how workload is distributed among those shipyards.

Deciding on a desired structure for the Australian shipbuilding industrial base is complex. Labor and overhead costs are certainly important factors, but national economic concerns may also play a role in deciding on the desired industrial base structure. Facility constraints must factor in, because the shipyards that currently build new ships are limited in the size of the vessels they can construct. Based on available information, the ASC South shipyard in Adelaide is the only one that has the facilities to assemble and deliver ships as large as the AWD and most likely the Future Frigate. If Australia envisions building large surface combatants in the future, ASC should be sustained in some capacity, or major facility upgrades will be needed at the other shipyards.

Planned workforce demand may be the most important factor when deciding on an industrial base structure. The demand will drop in a gap and increase with the start of a new program, suggesting the industrial base structure may change at various points in time. It will be difficult to sustain even small workforces during the gap between the end of the AWD program and the start of the Future Frigate program. Unless new work is added almost immediately (and it may already be too late), the block construction shipyards will start to shed workers, and their workforces will be dramatically reduced before the Future Frigate program starts. Some shipyards may even be forced to close unless options are pursued to lessen the gap. If a minimal number of workers are sustained before the start of the Future Frigate build, it will be difficult to spread those workers across multiple shipyards.[21]

There have been examples in which shipyards have started again after a period of closure (Cammell Laird and Appledore in the United Kingdom) and have built up a new construction workforce almost from scratch (Forgacs for the AWD program). During the workforce demand gap that Australia now faces, one strategy may be to focus

---

[21] Even with a fleet larger than RAN's, the United Kingdom has consolidated its shipbuilding to two shipyards—Govan and Scotstoun—having closed the third shipyard in Portsmouth. Huntington Ingalls Industries in the United States recently closed one of its three shipyards—Avondale in New Orleans.

on sustaining capabilities at one shipyard, with the plan of restarting shipbuilding capability at one or more shipyards when needed (which will come at a cost, as seen with the AWD program). The downside to sustaining one shipyard is the risk that some natural or man-made disaster will occur that results in the inability to build ships for some period of time.[22]

The various future acquisition options suggest a total demand for approximately 2,700 workers, possibly approaching 4,000 if the Future Frigate workload demand is as high as 7 million man-hours per ship. This number represents a medium or small shipyard by international standards. Sharing the rather low annual workforce demands among more than one shipyard may lead to inefficiencies in labor, excess costs in shipyard overhead, and scheduling problems. However, national decisions may suggest that two shipyards are preferred—one capable of building blocks and the second capable of building blocks, assembling them, and delivering a completed ship. This steady-state, two-shipyard structure would work best with a continuous build strategy.

Possible costs associated with a shared build strategy include:

- additional design effort to tailor the working drawings and three-dimensional product model to multiple shipyards
- need for additional government shipyard oversight
- facilitation to handle the construction and transport of blocks
- transportation between the various shipyards
- additional information technology upgrades at shipyards to support a common product model.

If the path of building complete ships in Australia is the preferred way forward for its shipbuilding, national-level decisions will be needed on how many shipyards to involve in building major surface combatants and how to distribute the workload. Based on our future projections of workforce demand, it may be costly to sustain more than two shipyards. The immediate problem is sustaining a shipbuilding workforce in the gap between the end of the AWD program and the

---

[22] It took several months for the Huntington Ingalls shipyard in New Orleans to recover after Hurricane Katrina.

start of Future Frigate construction. Finding options that can sustain more workers in the gap will provide more opportunities to use multiple shipyards.

Sustain a Limited Capability Australian Shipbuilding Industrial Base

As the previous analyses suggest, adequately sustaining a fully capable Australian shipbuilding industrial base will require changes to the construction plans of existing programs, including systematic drumbeats of new construction starts. A second future path for Australian shipbuilding is to follow a model recently used for the *Canberra*-class LHD program. The majority of the HM&E equipment for the LHD was built in Spain, with the final construction and outfitting of systems accomplished at the BAE shipyard in Williamstown, Victoria. A limited capability Australian shipbuilding industrial base would have the capacity to do some limited construction work but would concentrate on the installation and testing of the major combat and weapon systems.

Addressing the Short Term

The same issue of how best to sustain needed shipbuilding resources arises for the limited capability industrial base, but there are differences. First, fewer numbers of personnel are required for any new shipbuilding program. Second, there will also be a change in the number of various skills needed and the timing in demand for skills. In general, the demand for structural skills will decline to a greater degree than the need for outfitting skills. Finally, the length of the production gap will increase by approximately two to three years, because the first several years of production will occur overseas.

For the baseline analysis, the amount of labor is assumed to be 2.5 million fully productive man-hours, and follow-on ships follow a 95-percent unit learning curve. The build duration is assumed to be 18 quarters, as outfitting is the primary goal, with less emphasis on structural work and smaller amounts of labor required for management, support, and technical skills.[23] The planned delivery schedule mirrors the full-build scenario considered above (the full capability path); however, we assume that the first two years of production occur at a foreign shipyard. For example, given a 2020 start of production in the baseline case, work in Australian shipyards does not begin until 2022.

Figure 4.25 shows the shipbuilding workforce demand from today to the end of the Future Frigate program under the foregoing assumptions (final construction and outfitting of the Future Frigates performed in Australia). It also shows the number of workers for different sustainment levels during the gap, based on our estimated workforce demand profile. The figure shows a gap similar to that presented in Figure 4.2. However, the peak workforce demand is much less, because a good por-

Figure 4.25
Workforce Profile for Building Air Warfare Destroyers and Future Frigates (Limited Capability Path)

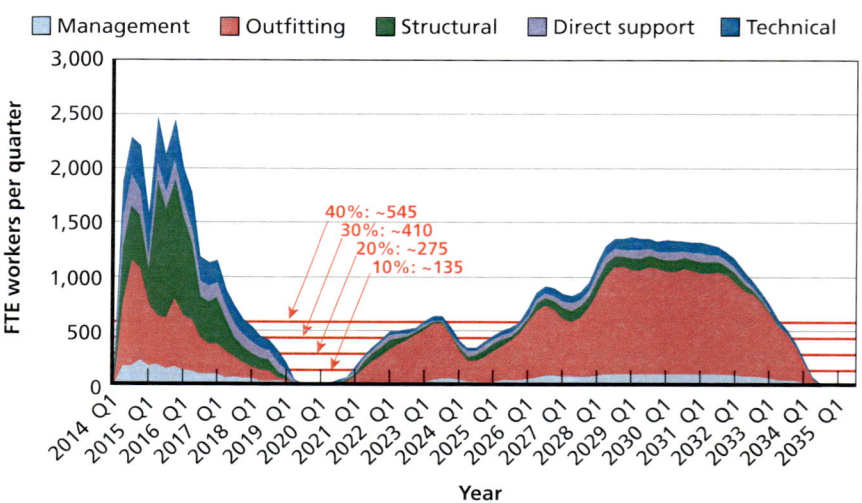

[23] We assume that the distribution of effort across skill categories approximates the distribution of effort exemplified by Australia's recent experience outfitting the LHDs. The workload demand function for the Future Frigates under the limited shipyard capability option will rise or fall depending on what tasks are assigned to Australian shipyards and what tasks are performed in a foreign shipyard. Schedules for starting work in Australia will also be affected by the workload distribution.

tion of the ship construction work is done in a foreign shipyard, and the gap extends for two additional years because of our assumption that the first two years of production will occur overseas.

Figure 4.26 shows the labor cost of sustaining different levels of experienced workers to meet the Future Frigate demand for the limited shipbuilding capability path. The shape of the curve is similar to that seen for the full shipbuilding capability path (see Figure 4.5), but the costs are lower due to the reduction in the Future Frigate demand function. And, as with the full shipbuilding capability path, sustaining an experienced workforce at the 20-percent level should provide a comparatively cost-effective way to meet future demands, after accounting for losses in productivity in the alternative case of not sustaining the workforce.

Figure 4.27 shows the unproductive man-hours as a function of the percentage of the peak workforce demand sustained during the gap. For example, if 20 percent of the peak demand is sustained, approximately 4 million unproductive man-hours will be included in the Future Frigate build. As expected, the number of unproductive man-hours drops

Figure 4.26
Total Labor Costs, by Workforce Sustainment Level (Limited Capability Path)

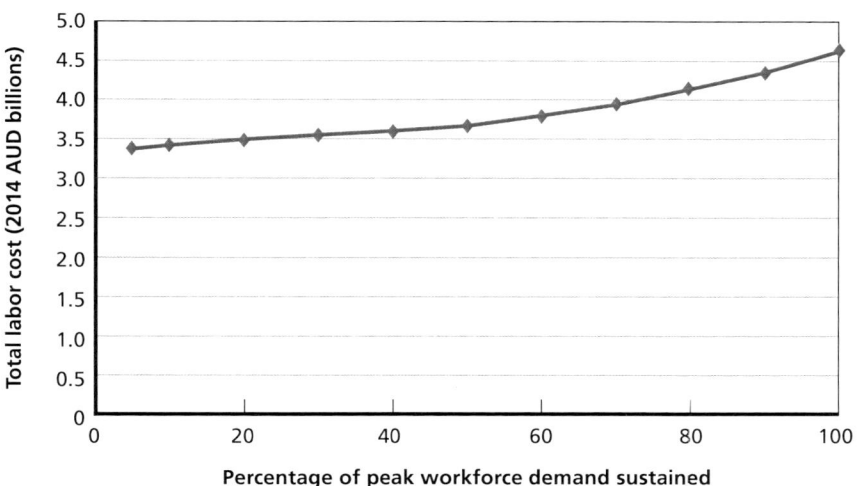

RAND RR1093-4.26

Figure 4.27
Average Cost per Full-Time-Equivalent Worker, by Workforce Sustainment Level (Limited Capability Path)

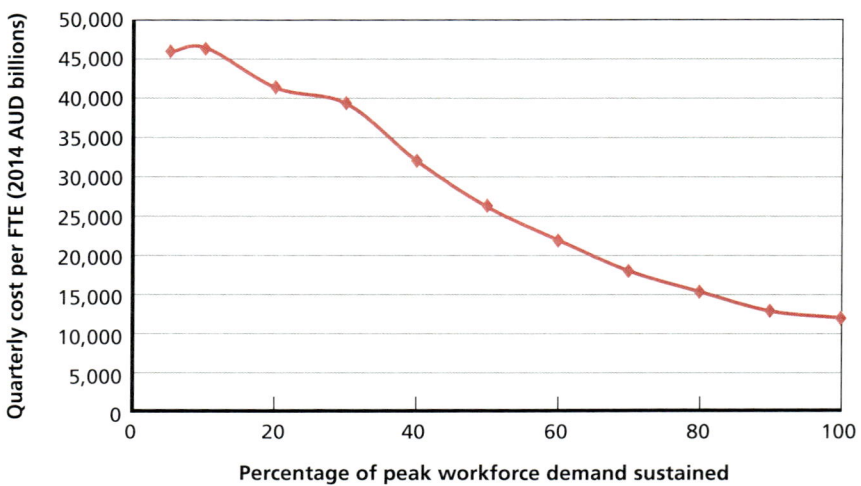

as larger percentages of the peak demand are sustained, because fewer new workers have to be hired and trained to full proficiency.

Figure 4.28 shows the schedule impacts of the limited capability path. Almost matching the base case for building new ships in Australia (see Figure 4.6), there is a significant schedule delay for the first several Future Frigates, either requiring the *Anzac* ships to stay in the force longer than planned or having a gap between the retirement of the first few *Anzac* ships and the in-service dates of the Future Frigates.

The options for lessening the gap between the end of the AWD program and the start of the Future Frigate program for the limited shipbuilding capability path for Australian shipbuilding are the same as those analyzed for the full capability option—move the start of the Future Frigate forward, insert a fourth AWD, or build patrol boats or OPVs at the major shipyards.[24] Table 4.6 summarizes the impacts of the various options on the labor costs and schedules.

---

[24] The patrol boats and OPVs would be completely built at an Australian shipyard, unlike large surface combatants, which are partially built overseas.

Figure 4.28
Schedule Implications of Sustaining 5 Percent of Peak Workforce Demand (Limited Capability Path)

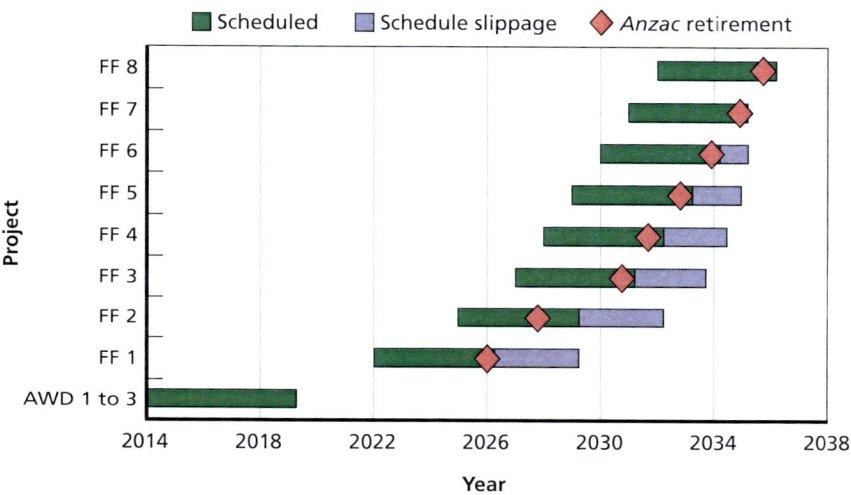

RAND RR1093-4.28

Table 4.6
Summary Labor Costs of Various Options for Workforce Sustainment (Limited Capability Path)

Option	Total Labor Costs (2014 AUD billions)	Total Schedule Delay Relative to *Anzac*-class Retirements (years)
Base case: 3 AWD, 8 FF (2020)	3.37[a,b]	16.50
3 AWD, 8 FF (2018)	3.34[a]	4.75
4 AWD, 7 FF (2020)	3.61[a]	1.75
4 AWD, 7 FF (2018)	3.59[a]	2.25
3 AWD, 8 FF (2020), 14 PB (2017)	3.45	17.75
3 AWD, 8 FF (2018), 14 PB (2017)	3.36	3.00

[a] Includes the cost of building 14 patrol boats starting in 2017 at different shipyards. We estimate the labor costs of the 14 patrol boats at AUD 120 million.

[b] Assumes a 5-percent workforce sustainment level in the gap period.

Table 4.6 shows the influence on total labor cost and delivery schedule of each of the discussed production plans. We see that starting the Future Frigate earlier reduces the total labor cost by a small margin, by increasing the efficiency of the workforce on a cost-per-FTE basis. However, the other gap-lessening options incur larger labor costs compared with the base case (which sustains a workforce level of 5 percent). In general, adding a fourth AWD increases total labor cost, as would be expected given the bigger ship.[25]

Adding patrol boats increases total labor cost slightly, because the workforce demand gap is two years longer and the patrol boats sustain structure skills more so than outfitting skills. That is, building patrol boats during the gap sustains skilled workers, who are not needed in large numbers for the Future Frigate in Path 2, while not sustaining the more important outfitting skills.

Logically, the option of adding OPVs would be less cost-effective in Path 2 for several reasons. First, the longer production gap in Path 2 means that several more OPVs would be needed to sustain a workforce for the duration of the gap. Second, the production of OPVs (like patrol boats) would emphasize structural skills much more so than outfitting skills, when the latter is what requires sustaining in Path 2. As a result, adding OPVs would be a costly way to lessen the gap, and we do not show results of model runs for these reasons.

Addressing the Longer Term

The problems with the longer-term workforce demand gap for the limited capability path (Path 2) are similar to the problems that confront building complete ships in Australia (Path 1)—what to do when the Future Frigate program ends and before the next major combatant program begins. The advantage of a partial Australian build is that the demand functions are lower. Some skills could be maintained in the gap after the Future Frigate by building smaller ships, such as patrol boats or OPVs. However, building patrol boats or OPVs will not help sustain the numbers and types of outfitting skills needed for the next

---

[25] We assume that the fourth AWD is a full build in Australia, unlike the Future Frigates, which are final construction and outfitting only.

major surface combatant. The drumbeat for the Future Frigate could be set at a new ship every two years to provide a continuous build strategy, or an additional three to five major combatants could be bought if the Future Frigate drumbeat is one ship per year. These ships could be additional assets for the RAN fleet or could be used to replace active ships that are retired before the end of their planned operational lives. The important issue when sustaining a skilled workforce in a gap between demands is ensuring that sufficient numbers of the right skills are sustained during the gap.

There is some overlap in the outfitting skills needed for ship support and the same skills for new ship construction. Depending on the timing and magnitude of future in-service ship support demands, there could be an overlap in the outfitting skills workforce between the segments of the naval ship industrial base.

Structure of a Limited Capability Australian Shipbuilding Industrial Base

As with an Australian industrial base that can fully build and deliver new ships, decisions are needed on how to structure a limited shipbuilding capability industrial base. The same advantages, disadvantages, costs, and risks arise with a structure that supports limited shipbuilding capability as were present with the full capability path. One difference for limited capability is that facilities to assemble and deliver large surface combatants are not needed, and other facilities besides ASC South, such as the BAE Williamstown shipyard, have the pier space to support a large ship during final construction and outfit. Workforce demand levels are reduced compared with the workforce needed to build complete ships, making it difficult to sustain multiple shipyards. At most, two shipyards could be sustained, but the future workload may be best used to sustain a single shipyard. However, maintaining a single shipyard does present the risk of losing capability if there is a natural or man-made disaster at that shipyard.

Sustain Only the Australian In-Service Ship Support Industrial Base

The third path open for the future Australian shipbuilding industrial base is to basically abandon new ship construction in Australia and buy as-built ships from other nations. Some specific system work may be accomplished in Australia, but the Australian ship-related resources and capabilities would be concentrated on supporting in-service ships rather than building new ones. As described in Chapter Two, sustaining a ship support industrial base is a function of the naval fleet and the policies for maintaining that fleet. Our initial analysis suggests that the current and future plans should adequately sustain an in-service ship support industrial base.

Summary

In this chapter, we have examined the labor costs of future paths, and options along those paths, for the Australian naval shipbuilding industrial base. We considered a future in which Australia has the capability and resources to build new major naval warships, one in which Australia sustains the ability to perform final construction and outfitting on ships where the basic HM&E portions are built overseas, and one in which the Australian naval ship industrial base buys fully outfitted ships from foreign shipbuilders.

Our initial analysis suggests that Australia could sustain a fully capable shipbuilding industrial base (Path 1) by taking some actions to mitigate a short-term gap in workforce demand between the end of the AWD program and the start of Future Frigate construction and by carefully managing a continuous build strategy in the longer term. Providing workforce demands in the short-term gap can sustain a skilled workforce for the buildup in demand from the Future Frigate program. In the first option, starting the build of the Future Frigate before 2020 would shorten the length of the gap. But there are numerous decisions yet to be made on the preferred acquisition path for the Future Frigate, and starting construction before 2020 is highly unlikely.

Building a fourth AWD largely mitigates schedule delays but requires significant funding for the extra ship (the purchase of the Aegis weapon system would be much larger than any cost savings in shipbuilding labor).

Starting to build new patrol boats in 2017 and having them built in the same shipyards that will build the Future Frigate can also help sustain a skilled workforce. The patrol boats do not mitigate the effects on schedule without also starting production of the Future Frigates before 2020.

Building OPVs in the gap can greatly reduce any delays in replacing the *Anzac*-class ships; moreover, the savings that arise from sustaining a productive workforce for employment by the Future Frigate would largely offset the labor cost of producing these additional ships. Bridging the short-term gap with three or more OPVs is an attractive option.

A continuous build strategy of every two years starting with the third Future Frigate will sustain a skilled workforce prepared to meet the demands of the next major warship program after the Future Frigate. However, building the last six Future Frigates with a drumbeat of two will result in the delay of new ships to replace the *Anzac*-class ships as they retire. This delay may cause the operational life of the *Anzac* ships to go beyond 30 years or present a shortfall in RAN major warships for several years. Another alternative for a continuous build strategy would be to keep the Future Frigate on a drumbeat of one and start to build the LMRVs as the Future Frigate build ends. But the LMRVs might best be viewed as a supplement to a Future Frigate drumbeat of two, providing a more effective bridge to the future major surface combatant program after the Future Frigate. In summary, there are several continuous build strategies, each with advantages and disadvantages that could sustain a cost-effective Australian shipbuilding industrial base.

The effect on workforce costs of the demand gap between the end of the AWD and the start of the Future Frigate construction is more muted for Path 2, in which Australia does not sustain the capability to build complete ships but rather maintains a limited shipbuilding capability that provides for the final construction and outfitting of ships built largely overseas. Both the total and peak workforce demand for the Future Frigates is reduced, because the majority of the basic HM&E equipment is built in foreign shipyards; however, the gap in Australian yards is extended by two or more years, because the first couple years of work will be completed overseas. Certain levels of various skills, especially final outfitting skills, should be sustained to meet the future demands. Starting to build the patrol boats or OPVs in 2017 can also help, although the patrol boats and OPVs will not sustain the types of final outfitting skills needed for the Future Frigate.

The longer-term strategy for sustaining the Australian naval shipbuilding industrial base with limited capability is similar to the strategy for the fully capable industrial base path. Building LMRVs will help, but a drumbeat of two will help sustain resources in the gap between the end of the Future Frigate construction and the start of the next major surface combatant for Australia.

The third path, forgoing any ability to build large naval warships, leads to a future in which only ship support is accomplished in Australia. As indicated in Chapter Two, this portion of the industrial base should be robust enough to meet future demands in a cost-effective manner, especially with the potential influx of workers leaving the shipbuilding industry.

In terms of the structure of the Australian shipbuilding industrial base that has the capability of building large surface combatants, our analyses suggest that it would be difficult and costly to sustain more than two shipyards during the short-term gap unless something is done to sustain production work at some level in that gap. Facility constraints also suggest that there is currently only one shipyard capable of assembling a large surface combatant. Relying on that one shipyard raises risks associated with the impact of possible natural or man-made disasters. The demands from a new surface combatant every two years, supplemented with builds of various smaller, less complex ships, will result in the need for a workforce of between 2,500 and 3,500 skilled workers. This may suggest two shipyards—one capable of building large blocks and the other capable of both building blocks and assembling those blocks into a completed ship. If such a two-shipyard structure is desired, more productive work may be needed in the gap to sustain the workforce at those two shipyards.

An Australian shipbuilding industrial base that focuses on the final build and outfit of ships largely built in other countries may suggest the need for only one shipyard, because workforce demands will be less than those for a complete shipbuilding program. A drumbeat of two would make it difficult to support two shipyards, because demands at any one of the shipyards would be spaced by four years.

CHAPTER FIVE

Benchmarking Australia's Naval Shipbuilding Industry with Comparable Overseas Producers

A key activity in RAND's analysis compared the relative performance of Australian industry with other naval shipbuilding nations. At the heart of this question is whether Australia pays a premium for its indigenously built naval vessels, and if so, what is the order of magnitude of that premium. This chapter discusses our initial findings with respect to the cost and schedule performance of Australia's naval shipbuilding. We focus exclusively on benchmarking outcomes (actual cost and schedule) rather than the practices (e.g., design-build process or outfitting and modularization practices) that drive this performance.

Benchmarking is the process of comparing the performance and practices of one firm, country, or system with another, at either an aggregate or unit level (e.g., program or item). This comparative process is used frequently in the commercial sector to identify strengths, weaknesses, and areas for improvement.[1] Benchmarking is often focused on identifying best practices and their degree of implementation across the comparison organizations. This research focused exclusively on performance benchmarking.

Performance benchmarking is used in the commercial shipbuilding world (for example, using metrics such as hours per compensated gross tonnage), because ships are generally single-purpose and built to an international standard. In this market, price is critical, so many of the metrics are focused on efficiency across the entire manufacturing process. However, with military shipbuilding, commonality is not a

---

[1] For more information about the benchmarking process, visit, for example, the website for the Construction Industry Institute.

given; low price does not always mean best value. Standards (particularly in the area of survivability) can vary greatly between countries, and warship capabilities can span a wide spectrum, even for similar mission sets and objects. For example, the capabilities of a frigate-sized ship can be quite different when performing various missions (e.g., antisubmarine warfare, strike, and surface warfare) than the capabilities of other warships of a similar size. The differences between the Italian and French frigate or frégate multimission (FREMM) multipurpose ships are a recent of example of this. For example, the Italian design can carry two helicopters but the French design carries one. On the other hand, the French design carries more capability in terms of land attack.

Quantitative naval ship benchmarking is hard to execute on a balanced basis due to the following reasons:

- the reluctance of countries and firms to provide such sensitive data on their vessels
- the difficulty in normalizing for these differing capabilities
- the fact that shipbuilders can be subsidized by the government or even government owned, so it is not clear that all the shipbuilding costs are captured
- the wide variability in exchange and inflation rates
- the fact that some countries have robust naval shipbuilding programs where others are more episodic
- not knowing whether the reported costs are all-inclusive (i.e., whether they include such items as recurring engineering, initial logistics support, and ordnance costs)
- differences in the definitions of even basic terms (e.g., the start of construction can mean very different things depending on the build approach employed).

Despite these differences, we will attempt to compare very basic cost and schedule outcome metrics for shipbuilding in Australia with several other countries. After that, we compare the total unit procurement price for shipbuilding, followed by comparing the time between nominal start of construction and commissioning.

Cost Benchmarking

Approach

Because of the risk in executing a naval cost benchmarking exercise and the uncertainty in the data, we will benchmark Australian naval shipbuilding relative to other countries using a three-method approach. The three approaches that we use are the following:[2]

1. *Input benchmarking.* The inputs to shipbuilding, such as labor, material, and equipment costs, are easily defined and are often captured by government economic organizations (e.g., the Australian Bureau of Statistics or the U.S. Bureau of Economic Analysis) or by private-sector firms to assist companies executing capital projects across the globe. For this type of benchmarking, RAND (1) gathered the public and private economic data relative to shipbuilding and other analogous industries (such as capital plant and offshore oil and gas construction), (2) developed a simple model of shipbuilding cost that uses these economic data as inputs, and (3) projected relative naval ship production costs for various countries.

2. *Comparative benchmarking.* For this type of benchmarking, we compare similar systems directly in terms of cost performance on a cost-per-metric-ton (CPT) basis. For example, one could compare the CPT of the F-105 built in Spain with the *Hobart*-class AWD. We broke the data into three broad types of vessels: frigates, destroyers, and amphibious ships. We performed adjustment and normalization of all the data in order to compare various ship classes within these broad types—that is, putting costs on a unit-procurement basis (a per-hull "purchase price," not including design) and converting costs to a uniform currency and year basis. For this approach, we identified appropriate comparative examples across several different countries: Australia, France, Japan, Republic of South Korea (Korea), Spain, the United Kingdom, and the United States.

---

[2] Qualitative benchmarking is a fourth benchmarking approach and generally focuses on business practices. Because this support research is quantitative in nature, it is excluded as a method for this study.

3. *Parametric benchmarking.* This type of benchmarking is a statistical method that defines a baseline (or typical) performance based on key system characteristics. The relationship is a parametric model. For example, for ships, one such parameter might be light displacement.[3] Using quantitative characteristics like these, one can define "average" cost using multivariate regression. The ratio of a country's performance on a ship to the average performance based on the model will then define a performance metric relative to industry average—higher than one meaning more expensive.[4] However, to execute such an approach, the costs and technical characteristics must be gathered on many types of vessels for each country of interest. We explore one such regression using the same data used for the comparative benchmarking.

By employing these multiple benchmarking approaches, we estimate the cost performance of Australia relative to other nations and identify the uncertainty in the performance characterization. All the methods are subject to a particular issue—not being able to control for all important factors (specification error or omitted variables in a statistical sense). Thus, the consistency in the answers for each method will indicate the level of confidence one can have in them. If the results are widely disparate, then the results should be viewed with caution. If the results between the methods are similar, then the results can be viewed with more confidence.

Caveats

We caution the reader not to overinterpret any one particular cost value. We report values to a precision consistent with the original sources. However, the precision does not always imply the accuracy of

---

[3] For an example of this approach, see Mark V. Arena, Irv Blickstein, Obaid Younossi, and Clifford A. Grammich, *Why Have the Cost of Navy Ships Risen? A Macroscopic Examination of the Trends in U.S. Naval Ship Costs over the Past Several Decades*, Santa Monica, Calif.: RAND Corporation, MG-484-NAVY, 2006.

[4] In statistical terms, this would be related to the average residuals for that country's observations in the regression model.

the values. This is particularly true for the reported ship costs that one finds in the press or on the Internet. While we have made an effort to use authoritative sources and check for completeness, information on some countries, such as Korea, is difficult to obtain. Other times, the sources of data may be inconsistent. For these reasons, we will report cost metrics (where possible) in a range (i.e., low, mid, and high; or low and high) to reflect, in part, this uncertainty. The ranges do not imply a confidence interval but rather a range of potential values due to the sensitivity in inputs (such as the annual variability in the currency exchange rate); the reported range of the data, or the range from different sources of the same data; or alternative assumptions with respect to the data (e.g., corrections for nonrecurring engineering costs included but not specified). In the section on comparative benchmarking, we further discuss some of the data limitations of the example ship prices used. The same caveats apply to the parametric analysis because we used the same data.

Input Benchmarking

Input benchmarking examines the relative costs of producing the end item—in this case, a naval warship. Shipbuilding generally comprises three different sources of cost: labor, material, and equipment.[5] It is challenging to find these costs specific to naval shipbuilding across multiple countries.[6] In fact, there are no reported public statistics that are specific to naval construction. However, some countries track and report their wage costs for boat and shipbuilding more broadly. In Table 5.1, we list the recent direct hourly wage metrics for Australia (in Australian dollars, or AUD), the United States (in U.S. dollars, or USD), and the United Kingdom (in pounds). The table displays three different values for each country: the average direct hourly pay for each worker, that hourly rate converted to an Australian dollar, and

---

[5] Obviously, the costs for shipbuilding are much more nuanced in terms of their variety and type. But as we need to compare across multiple countries, it is only feasible to compare at this aggregate level.

[6] We are limited in what data RAND may possess from other research due to its proprietary nature. Moreover, there are limits on what we can share of U.S. and U.K. data that are not open source.

Table 5.1
Direct Hourly Wage Rates for Boat and Shipbuilding

Country	Direct Pay per Hour	Converted Direct Pay (AUD per Hour)	Relative Pay (U.S. = 1.0)	Source
Australia	AUD 38.80[a]	38.80	139%	Australian Bureau of Statistics, "Employee Earnings and Hours, Australia," May 2013
U.S.	USD 24.50	27.84	100%	U.S. Bureau of Labor Statistics, "National Industry-Specific Occupational Employment and Wage Estimates: NAICS 336600—Ship and Boat Building," May 2013b
U.K.	£16.35	29.75	107%	U.K. Office for National Statistics, "Weekly Pay—Gross (£)—For Full-Time Employee Jobs: United Kingdom, SIC2007, Table 16.1a," 2013

NOTE: Values are reported on a fixed 2013 basis.
[a] Value has been escalated from 2012 to 2013 to be on a comparable basis.

the wage rate relative to a U.S. basis.[7] It also displays the source of the data. From the relative pay column, we can see that Australian direct pay rates are approximately 40 percent higher than U.S. rates and 30 percent higher than U.K. rates. So, if labor cost dictated relative naval shipbuilding prices (and they are a substantial portion of those costs), then one would expect that ship prices in Australia would be 20 percent to 30 percent higher than a U.S. or U.K. basis. However, these costs are just one part of the shipbuilding value chain and are not the full labor price that includes burdens and profit. They should be viewed as indicative of relative costs only.

To broaden the list of countries, we needed to choose higher-level business sectors. One such sector is manufacturing. Table 5.2 shows the average hourly compensation costs for 2012, in U.S. dollars. Unlike before, these labor costs include some of the indirect labor costs, such as sick pay, vacation, health insurance, unemployment insurance, and

---

[7] We have chosen a U.S. basis because the majority of the other input data that are shown in this chapter are on a U.S. basis.

Table 5.2
Hourly Compensation Costs in Manufacturing (2012)

Country	USD per Hour	Index (U.S. = 1.0)
Denmark	48.47	1.36
Australia	47.68	1.34
United States	35.67	1.00
Japan	35.34	0.99
Italy	34.18	0.96
United Kingdom	31.23	0.88
Spain	26.83	0.75
Korea	20.72	0.58
Estonia	10.41	0.29

SOURCE: U.S. Bureau of Labor Statistics, "International Comparisons of Hourly Compensation Costs in Manufacturing," May 2013a.

NOTE: Rates are reported on a U.S. dollar basis to be consistent with the source. Compensation costs include direct pay, social insurance expenditures, and labor-related taxes.

payroll taxes (but not fees or profits). Similar to the results in Table 5.1, Australian labor costs are approximately 35 percent higher than a U.S. basis. Moreover, only Denmark is more expensive in the comparison country list. However, the recent Danish *Iver Huitfeldt* program used a mix of Danish and Estonian labor for construction. With the Estonian labor rate being much lower than any other country on the list, this offsets the high Danish labor rates. Spain and Korea are nearly half that of Australia. Again, this result suggests that Australian naval shipbuilding costs will be significantly higher than most other comparator countries.

Perhaps a better industrial sector to compare with naval shipbuilding is construction in the oil, chemical, and gas industry—both on and offshore. This industry uses many of the same skills as naval shipbuilding. As part of our discussions with the naval repair organizations in Australia, several indicated that they use the same subcontractors for naval repair work as the construction industry for oil, chemical, and gas. As that construction industry is highly competitive, it is not surprising that several firms benchmark construction cost. One of the leaders in this area is Compass International Inc., which publishes annual reports on the construction costs for many countries.[8] We draw upon the data published in the *2014 Global Construction Costs Yearbook* to examine relative input costs across a wide range of construction types.[9] Because of the copyright for that material, we cannot reproduce the raw data contained within that report. However, we can form a composite index that weights labor, material, and equipment prices for each comparator country. We will assume a split of labor, equipment, and material of 40 percent, 30 percent, and 30 percent, respectively, which is roughly typical of naval ships (but can vary considerably based on ship type). Figure 5.1 shows a composite index for the same comparator countries as Table 5.2 (except Estonia). For each country, we show a high value (top of the blue bar), medium value (red line), and low value (bottom of the blue bar) to indicate the variability in the data.

The results shown in Figure 5.1 indicate that Australia is significantly more expensive than most of the comparator countries, with the exception of Denmark. Korea is the least expensive, at roughly 10 percent below a U.S. basis. Italy, Japan, and Spain are also below the U.S. value. Denmark, Australia, and the United Kingdom are higher.

A limitation of the above analysis is that it assumes labor productivity commensurate with construction and not shipbuilding. First Marine International has produced several reports examining shipyard productivity by measuring the hours per compensated gross tonnage (CGT).[10] These reports observe that the U.S. major naval shipyards have a productivity of 30 to 60 hours per CGT.[11] They also report

---

[8] See the Compass International website for more information.

[9] Compass International Inc., *2014 Global Construction Costs Yearbook*, 2014.

[10] See, for example, First Marine International Ltd., *First Marine International Findings for the Global Shipbuilding Industrial Base Benchmarking Study*, Part 2: *Mid-Tier Shipyards, Final Redacted Report*, February 6, 2007. Compensated gross tonnage is a method of comparing the weights of different types of commercial ships with one another by using adjustment factors that depend on the ship type (e.g., tanker, dry cargo, ferry). For more information, see Organisation for Economic Co-operation and Development, Directorate for Science, Technology, and Industry, "Compensated Gross Ton (CGT) System," 2007.

[11] First Marine International, 2007, Figure 5.2.

Figure 5.1
Relative Oil, Chemical, and Gas Plant Construction Costs

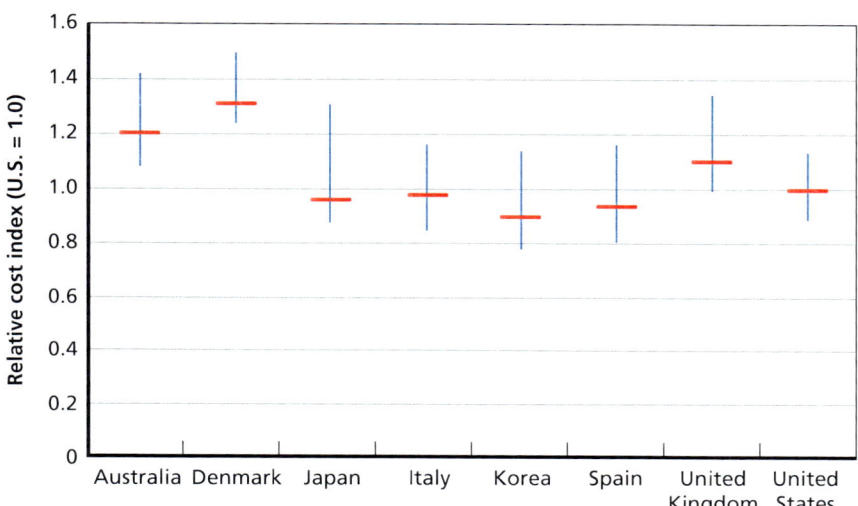

SOURCE: Authors' calculations based on Compass International, Inc., 2014.
RAND RR1093-5.1

ranges for European, Korean, and Japanese large yards. We can use those historical ranges—albeit dated—as relative shipbuilding labor productivity values.

While there have been reports of similar First Marine International benchmarking for Australia, none of it has been made public due to its sensitivity. There is one public statement that the Australian AWD program expects to achieve overall 80 hours per CGT (despite the first ship being significantly higher at 150 hours per CGT).[12] The *Future Submarine Industrial Skills Plan* also anchors Australian productivity at 80 hours per CGT (again, the anticipated target productivity for the AWD) with a high-low range from 50 to 110 hours per CGT.[13] We assume that average performance for Australia is in a

---

[12] Commonwealth of Australia, *Future of Australia's Naval Shipbuilding Industry: Tender Process for the Navy's New Supply Ships*, Part I, Economics References Committee, August 2014b.

[13] Commonwealth of Australia, 2013b.

range of 80 hours (±33 percent) per CGT based on that testimony, the *Future Submarine Industrial Skills Plan*, and the assumption that that Australian shipbuilding carries its learning from the AWD program.[14] Our productivity assumption and the assumption in the submarine skills report are nearly identical. Figure 5.2 shows the revised relative cost values using these labor productivity values. Of the comparator countries, Australia is the highest at roughly 45 percent more than a U.S. basis. Korea, Japan, and Spain are significantly less than a U.S. basis—roughly 20 percent less. Italy and Denmark/Estonia are

Figure 5.2
Relative Construction Costs, Based on First Marine International Shipbuilding Productivity

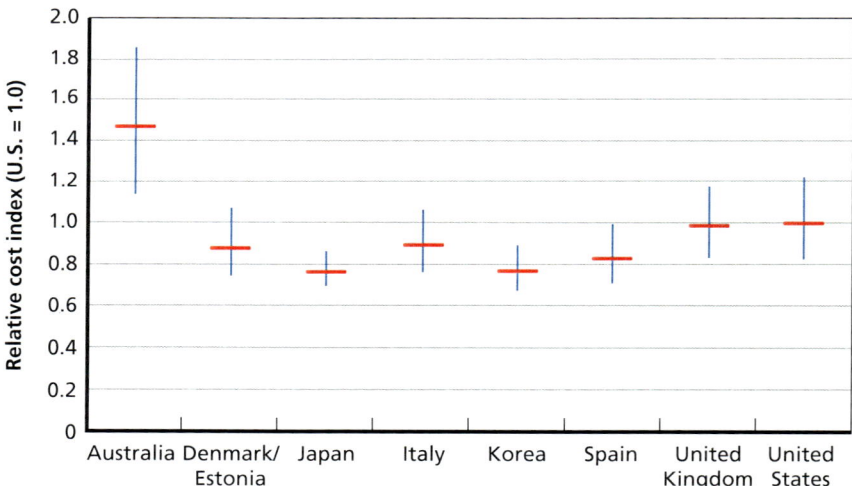

SOURCE: Authors' calculations based on Compass International, Inc., 2014; Commonwealth of Australia, 2013b; and First Marine International Ltd., 2007.
RAND RR1093-5.2

---

[14] We assume a range of average performance hours, because this value is uncertain and highly debated. However, one should not overly focus on only labor productivity when considering a value proposition of building domestically or not. Relative labor rates come into play. Furthermore, labor costs generally account for only 40 percent of a combatant's price. In the next two sections, we benchmark at the total price level and arrive at similar results. Thus, this assumption on productivity is reasonable.

approximately 10 percent,[15] and the United Kingdom is roughly the same as the U.S. basis.

The relative results for Australia might seem unreasonably high. However, they are based on the productivity stated for the AWD program and consistent with the view of that program's performance.[16] In the next section, we compare the AWD on a USD-per-metric-ton basis and arrive at similar results (about 40 percent higher). Moreover, the other input benchmarks suggest that naval shipbuilding costs in Australia for combatants are approximately 20 percent to 40 percent higher than a U.S. basis.

Comparative Benchmarking

In this section, we compare the unit procurement costs for several naval vessels built by different countries. We group these examples by three different vessel types to aid in the comparisons: frigates, destroyers, and amphibious vessels. Table 5.3 lists some basic physical characteristics for the frigates, as well as the number of hulls produced. Tables 5.4 and 5.5 provide similar information for the example destroyers and amphibious vessels, respectively.

In Tables 5.6 through 5.8, we present the source of the cost data for each ship. Note that except for the Spanish and Korean ships, the costs derive from authoritative government sources.

---

[15] Assumes an 80:20 labor split between Estonia and Denmark.

[16] Commonwealth of Australia, 2013b. See also Australian National Audit Office, 2013, p. 260: "By November 2013, it was costing ASC $1.60 to produce work that was originally estimated to cost $1.00." This implies a lower realized productivity by nearly 40 percent than assumed in the original estimate baseline (which would increase the man-hours by 60 percent from that planned). Our 80 hours per CGT assumed value compared with the midpoint of the U.S. range would suggest an approximate 80-percent labor hour premium relative to the U.S. norms.

Table 5.3
Comparison of Physical Characteristics and Hulls Produced, Frigates

Ship	Country	Length (m)	Beam (m)	Draft (m)	Full Load Displacement (metric tons)	Number of Hulls Produced
F590 FREMM	Italy	144.0	20.0	4.5	6,500	10
D650 FREMM	France	142.2	19.7	5.4	6,096	11
De Zeven Provinciën air defense and command frigate (LCF)	Netherlands	144.2	17.0	7.0	6,050–6,145	4
Iver Huitfeldt	Denmark	138.7	19.8	6.3	6,645	3
Anzac	Australia	118.0	14.8	4.4	3,600	10[a]
Incheon	Korea	114.0	14.0	4.0	3,251	6[b]
Oliver Hazard Perry FFG-7	United States	136.0	14.0	6.7	4,166	51[c]
Littoral combat ship (LCS) Freedom variant	United States	118.8	17.6	3.9	3,354	2[d]
LCS Independence variant	United States	127.6	31.6	4.4	2,841	2[d]

SOURCES: Unpublished RAND research on Future Frigate design and construction; IHS, undated.

[a] Includes two ships for New Zealand.

[b] First batch, 18–24 planned.

[c] U.S. production only.

[d] Two of each variant completed, more in production.

Table 5.4
Comparison of Physical Characteristics and Hulls Produced, Destroyers

Ship	Country	Length (m)	Beam (m)	Draft (m)	Full Load Displacement (metric tons)	Number of Hulls Produced
Japan Defense Ship (JDS) *Akizuki* (DD)	Japan	150.5	18.3	5.3	6,800	4
JDS *Atago* guided missile destroyer (DDG)	Japan	164.9	21.0	6.2	10,323	2
Hobart (AWD)	Australia	146.7	18.6	7.2	7,000	3
Arleigh Burke Flight IIA (DDG)	United States	155.3	20.3	9.3	9,515	62[a]
Sejong Daewang Korean Destroyer Experimental (KDX)-3	Korea	165.9	21.4	10.5	10,455	3[b]
Daring Type 45	United Kingdom	152.4	21.2	5.3	7,570	6
Cristóbal Colón F105	Spain	146.7	18.6	4.8	6,350	1[c]

SOURCES: Unpublished RAND research on Future Frigate design and construction; IHS, undated.

[a] Completed, more planned.

[b] Completed, more potential.

[c] Preceded by four ships of similar design.

Table 5.6
Sources of Cost Data, Frigates

Ship	Source of Cost Data
F590 FREMM	Italian Ministry of Defense, *Documento Programmatico Pluriennale per la Difesa per il Triennio 2013–2015*, April 2013; prior RAND research[a]
D650 FREMM	Daniel Reiner, Xavier Pintat, and Jacques Gautier, *Défense: Équipement Des Forces et Excellence Technologique Des Industries De Défense*, Senate Presentation, November 21, 2013; prior RAND research[a]
De Zeven Provinciën LCF	Prior RAND research[a]
Iver Huitfeldt	Private correspondence between RAND and Odense Maritime Technology (OMT); prior RAND research[a]
Anzac	Budget data provided by the Australian White Paper team[a]
Incheon	Defense Industry Daily, "Korea's New Coastal Frigates: The FFX Incheon Class," August 25, 2014
Oliver Hazard Perry FFG-7	U.S. budget documents (NAVSEA017, P-22 Reports Database, October 12, 1995)[a]
LCS	U.S. Department of the Navy, *Department of the Navy Fiscal Year (FY) 2015 Budget Estimates: Justification of Estimates—Shipbuilding and Conversion, Navy*, March 2014

[a] Not publicly available.

represent the first three ships after restarting the production line. The LCS costs represent the FY 2011 and FY 2012 procurements. The LPD-17 cost range is for the FY 2009 and FY 2012 authorizations. The LHA-R is the FY 2011 authorization (one-off restart hull).

2. Convert from local currency to a U.S. dollar basis at either the midpoint spending or the year basis for the local currency, depending on how the costs were reported.
3. Escalate to a common-year basis for the U.S. dollar (2014) using the SCN (SAP Community Network) index, which is the U.S. government's official escalation rate for ship procurements.[18]
4. Calculate a 2014 USD-per-metric-ton factor using full displacement. Ideally, it would be better to use *light ship displacement*

---

[18] Naval Center for Cost Analysis, "NCCA Inflation Indices and Joint Inflation Calculator," March 2014.

Table 5.7
Sources of Cost Data, Destroyers

Ship	Source of Cost Data
JDS *Akizuki*	Japanese Ministry of Defense, *Defense Programs and Budget of Japan: Overview of FY2014 Budget*, December 2013
JDS *Atago*	Bureau of Finance and Equipment, *Current Situation of Ship Production and Skill Base*, Japanese Ministry of Defense, March 2011
Hobart (AWD)	Australian National Audit Office, 2013; budget data provided by Australian White Paper team[a]
Arleigh Burke Flight IIA	U.S. Department of the Navy, *Department of the Navy Fiscal Year (FY) 2013 Budget Estimates: Justification of Estimates, Shipbuilding and Conversion, Navy*, February 2012
Sejong Daewang KDX-3	Defense Industry Daily, "Korea's KDX-III AEGIS Destroyers," May 27, 2014
Daring Type 45	U.K. National Audit Office, *The Major Projects Report 2011*, U.K. Ministry of Defence, November 16, 2011, appendixes and project summary sheets
Cristóbal Colón F105	Infodenfensa.com, "Special Weapons Programs Recorded a Deviation of 32% Cost," December 10, 2011[a]

[a] Not publicly available.

Table 5.8
Sources of Cost Data, Amphibious Vessels

Ship	Source of Cost Data
JDS *Izumo*	Japanese Ministry of Defense, *Defense Programs and Budget of Japan: Overview of FY2010 Budget*, 2009
Canberra LHD	Australian National Audit Office, 2013; budget data provided by White Paper team[a]
Juan Carlos (L61)	Naval Technology, "Juan Carlos I Landing Helicopter Dock, Spain," web page, undated(b); Spanish Ministry of Defense, *Evaluación de los Programas Especiales de Armamento (PEAs)*, Madrid: Grupo Atenea, September 2011
San Antonio-class LPD-17	U.S. Department of the Navy, 2014
America-class LHA-R	U.S. Department of the Navy, 2014
Wasp-class LHD-1	Congressional Budget Office, *Total Quantities and Unit Procurement Cost Tables: 1974–1995*, Publication 18099, April 13, 1994
Albion-class LPD-R	U.K. National Audit Office, *The Major Projects Report 2000*, U.K. Ministry of Defence, November 22, 2000, Appendix 3

[a] Not publicly available.

(ship weight not including variable loads, such as fuel ordnance and supplies).[19] But these weight data were not available for all examples.
5. Calculate a relative cost index based on dividing the 2014 USD-per-metric-ton value by a U.S. ship example. For the frigates, the normalization basis was the FFG-7 class; for the destroyers, it was the DDG-51 class; and for the amphibious ships, it was the LPD-17.

There are a handful of important limitations and uncertainties embedded in this analysis of which the reader needs to be aware.

- *Comparability of the costs.* The quality of the production cost data is highly variable. Where possible, we have attempted to get data from official government sources, such as defense departments, government budget agencies, and audit agencies, or data from prior RAND research. However, these cost data are not always presented on a comparable basis. For example, some sources report costs in detail, so it is possible to isolate the total production cost on a per-hull basis. Other sources, however, only report an average unit production price. Still other sources report costs as total program price.
- *Inclusiveness of the costs.* Another concern is whether all the procurement costs are included. For example, do the prices include government-furnished material and equipment costs? Most of the data are relatively complete, but for a few ship classes (e.g., the Japanese, Korean, and Spanish examples), we do not know whether all these costs are captured or whether the quoted prices represent a "contract" value with the shipbuilders.
- *Currency and escalation.* One of the difficulties in benchmarking international programs is placing them on the same currency basis and removing the effects of inflation. There is no simple process to do this. For our analysis, we translated the local currency to a fixed U.S. dollar basis at the midpoint of the program spending. We chose a U.S. dollar basis because RAND has the

---

[19] See Arena et al., 2006.

best exchange rate and escalation data to control variations. Alternatively, we explored escalating local currency with a local escalation index to a 2014 basis and then converting to U.S. dollars. Overall, there were small differences in the results between the methods. The one exception was with the older Australian ships (e.g., *Anzac* frigates) because the Australian dollar has greatly appreciated versus the U.S. dollar in recent years. We chose the former conversion-escalation method, as it seemed to yield the most consistent and reasonable results. If we had chosen the alternative method, the *Anzac* class would seem much more expensive (relatively). Also, we are interested in the purchasing power at the time of build, not at some future time.

- *Unequal effectiveness.* We normalize for overall size differences between ships by examining relative costs on a per-metric-ton basis. However, we have not adjusted for military capability or effectiveness, because that is beyond the scope of this analysis. Our comparisons are purely based on cost, not on cost-effectiveness.
- *Government support and investment in shipbuilding.* One factor we cannot control for is the local government's amount of investment and cost for the naval shipbuilding enterprise that does not appear in the ship prices. All countries make some investments in their shipbuilding enterprises. However, some have had either direct partial ownership or some control of that enterprise (e.g., France). So, whether all of the support, management, and investment costs are fully reflected in the ship prices is unknown.
- *Domestic budget prices versus foreign offer prices.* An important caveat on the price is that it is the price to a particular government and not the price (or relative price) that the Australian government could obtain the ship as an off-the-shelf solution. For one, any design would have to be modified to the Australian needs and regulations. Furthermore, the offer price could be higher or lower depending on other demands of that industry. We cannot forecast the health of any particular naval shipbuilding market nor how aggressive any particular firm might be toward a business opportunity. So, our analysis will be based on the embedded profit and not make other adjustments for market conditions now or in the future.

Because of these uncertainties, the comparative cost values should be viewed as indicative only and not precise relative costs.

In Tables 5.9, 5.10, and 5.11, we show the relative cost index by ship type. Where we have multiple values, we show a low-high range for the results. Where we have only one, we display that as the low value. Note that even though ships may have a single average unit procurement cost, they will still have a low-high index range due to variability in the exchange rate. Another point of clarification is that the costs shown in these tables are *unadjusted* for any productivity or labor cost differences between the countries. These tables are meant to show the relative building costs between countries and not priced as built in Australia (except for the Australian examples).

Table 5.9
Unit Procurement Cost and Relative Index Cost Data, Frigates

Ship	Country	Relative CPT Index[a]	
		Low	High
F590 FREMM	Italy	0.95	1.00
D650 FREMM	France	1.18	1.24
De Zeven Provinciën LCF	Netherlands	1.00	1.07
Iver Huitfeldt	Denmark	0.56	0.62
Anzac	Australia	1.36	1.48
Incheon	Korea	0.65	0.75
Oliver Hazard Perry FFG-7	United States	0.93	1.07
LCS[b]	United States	1.42	1.44

SOURCES: See Table 5.6.
[a] *Oliver Hazard Perry* FFG-7 class is set to 1.0.
[b] We do not have costs split by variants, so we report an average cost instead. Also, these costs do not include mission module costs.

Table 5.10
Unit Procurement Cost and Relative Index Cost Data, Destroyers

Ship	Country	Relative CPT Index[a]	
		Low	High
JDS *Akizuki*	Japan	0.46	0.51
JDS *Atago*	Japan	0.76	0.80
Hobart AWD	Australia	1.24	1.39
Arleigh Burke Flight IIA	United States	0.86	1.14
Sejong Daewang KDX-3	Korea	0.54	0.56
Daring Type 45	United Kingdom	0.89	1.05
Cristóbal Colón F105	Spain	0.91	0.96

SOURCES: See Table 5.7.
[a] *Arleigh Burke* Flight IIA class set to 1.0.

Overall, the performance for Australian naval shipbuilding is mixed based on the example programs in Tables 5.9 through 5.11. For the *Anzac* frigate program, the cost index was roughly 40 percent more than the U.S. FFG-7 cost basis. This relative difference is similar to the results we saw for the cost index in Table 5.2. The difference is driven mainly by high labor costs and lower productivity. Not surprisingly, the Korean frigate is the least expensive—around 30 percent less than the U.S. basis and a bit lower compared with the earlier values from Figure 5.2. However, this average unit cost value is very uncertain because it was based on a second-hand report and not an official source.[20] The Danish frigate is quite low relative to the other ships. This ship was built and procured in a commercial-like manner (e.g., building practices and HM&E equipment that are commercial) but with survivability features added (e.g., shock and ballistic protection). The ships were also built in productive and automated commercial ship-

---

[20] For more information, see Defense Industry Daily, 2014.

Table 5.11
Unit Procurement Cost and Relative Index Cost Data, Amphibious Vessels

Ship	Country	Relative CPT Index[a]	
		Low	High
JDS *Izumo*	Japan	0.86	1.01
Canberra LHD	Australia	0.99	1.25
Juan Carlos (L61)	Spain	0.50[b]	0.64[b]
San Antonio-class LPD-17	United States	1.64	1.68
America-class LHA-R	United States	1.55	
Wasp-class LHD-1	United States	0.92	1.08
Albion-class LPD-R	United Kingdom	0.59	0.65

SOURCES: See Table 5.8.
[a] *Wasp* class set to 1.0.
[b] Per discussions with the White Paper team and the *Canberra* program office, costs likely only represent build-phase costs.

yards. There were some questions during the initial draft stage of this report on whether the total procurement for the *Iver Huitfeldt* class fully included all the comparable procurement costs. However, the Danish Navy and OMT provided cost details that included the cost of integration work done by the navy and the cost of reused equipment; so, we are reasonably confident these costs are on a comparable basis. The LCS is the highest on a CPT basis (mid-point), and it would be even higher had we included the mission modules as part of the unit cost. As we will see in the next section, this high value is partially driven by the high maximum speed for the vessels.

The destroyer examples should be the most comparable, because all but the Type 45 and Japan's *Akizuki* use an Aegis radar and combat system. Here, the Australian example is the most expensive, around 35 percent more than the U.S. baseline. And these costs are *low estimates* for the *Hobart* class, because the program is updating its cost baseline currently. The difference is consistent with the difference observed in Figure 5.2. The Korean and Japanese vessels are roughly 25 percent to 50 percent less than the U.S. baseline. However, we have the same concerns about the completeness of the data as we did for the Korean frigate. We are not certain that all costs are captured. Spain and the United Kingdom are slightly below the U.S. baseline value.

Australian naval ship costs compare better in the amphibious ship example. The Australian LHD was about 12 percent higher than the U.S. *Wasp* class—the closest U.S. comparator. The Spanish *Juan Carlos* (the parent design for the *Canberra* class) is the least expensive—about 45 percent less than the U.S. baseline. The fact that the Spanish design is nearly half the value of the Australian one, despite being very similar ships, appears inconsistent. However, there is some question about the inclusiveness of the Spanish costs, and discussions between the White Paper team and the Australian LHD program office suggest that Spanish LHD costs include only the direct shipbuilding costs and not such other costs as logistics support and program management. If we compare the *Canberra* class on a similar basis, the *Canberra* index would be 0.69—about 8 percent higher than the *Juan Carlos*. This difference seems more reasonable. A significant portion of the Australian LHD is being built in Spain, so one should not infer that the Australian LHD costs fully represent Australian naval shipbuilding performance.

As a caution, one should note that there is a much broader range in capability and missions for the amphibious ships than for the frigate and destroyer comparisons. For example, the *San Antonio* class carries more weapon systems than do other examples in the table (for example, it has extensive combat data systems), which partially explains its higher relative CPT. Also, using full load displacement to normalize cost is more inaccurate for the amphibious ships because they have significant cargo and materiel-carrying capabilities. The full displacement includes this cargo weight, so amphibious ships with larger cargo values might seem less expensive on a CPT basis because the cargo is not part of the ship costs. So, the results of the amphibious comparisons should be viewed with more caution.

Parametric Analysis

In the previous section, we explored a basic CPT metric as a measure of relative cost for three different ship types. One might wonder how representative a simple CPT metric is. For example, there might be scaling efficiencies or complexity difference that might make this metric misleading. Moreover, ship cost is affected by more than just ship weight. To try to address such questions, we will use a parametric approach through multivariate regression analysis. Note that the same caveats in terms of data quality also apply to this parametric analysis.

In the past, we have successfully used this parametric approach to determine relative ship costs.[21] For this analysis, we are limited to ship parameters that are publicly available for the foreign ships. So, using such parameters as light ship weight or power and system density will not be feasible because such information is generally not made public. Based on the small sample size—20 ships in Tables 5.3, 5.4, and 5.5—we will not be able to use more than two or three parameters. The best regression we found was a log-log relationship based on ship weight and flank speed. Also, we included a binary term to reflect those data points for which we had less confidence in their completeness.

This simple formulation results equation explains three quarters of the total variance. It is worth discussing the values of the coefficients. The coefficient for the weight term is very nearly one. This indicates that cost scales linearly with weight and validates our earlier use of a CPT metric in the comparative analysis. The coefficient for speed is nearly three, which indicates that ship cost and maximum speed have a cubic relationship. Interestingly, power and ship speed also have a cubic relationship. The "uncertain" term was introduced into the regression analysis to see whether the cost data that were less certain (e.g., the Japanese, Korean, and Spanish ship costs identified earlier) were different. They were much lower; what we cannot say is whether this is an intrinsic difference (they do produce ships for lower costs) or whether the costs are not complete. Our earlier productivity and labor analysis suggests that they are lower, however. The term is used so that the regression results are not potentially skewed by these observations.

---

[21] Arena et al., 2006.

More interesting for this study is the resulting residuals based on the regression equation. The *residuals* are the unexplained variance of the observations. For our purposes, they should relate, in part, to the differences in cost between countries. If we t-test the residuals against a binary term for whether or not the ship is an Australian one, we should get an idea of the cost premium. This statistical test indicates that the coefficient for Australian ships is 0.28, with a standard error of 0.19. The probability that this coefficient is insignificant (not statistically meaningful) is less than 6 percent. This coefficient translates as a 32-percent premium for Australian ships, with a standard range of 9 percent to 60 percent. This premium is consistent with our earlier results.

Influence of Exchange Rate

An important consideration in any domestic-versus-overseas build decision lies with exchange rate values and the relative purchasing power of the Australian dollar. When the Australian dollar is strong, overseas builds are more competitive, but when it is weak, they are less competitive. As an example, let's look at the Australian price to buy a DDG-51-class destroyer in two different time periods, 2000 and 2014. First, assume that the United States will sell it to the Australians at the same price they paid. Next, we convert the cost in U.S. dollars to Australian dollars using the exchange rate at that time. Finally, we inflate the 2000 purchase in Australian dollars to 2014 prices using an Australian inflation index. Table 5.12 shows the sequence in numbers. We evaluate to different 2014 exchange rates to show the volatility in the costs—a July and November average. It turns out that buying a DDG using a July 2014 exchange rate would be about 14 percent *less* than buying the same nominal ship in 2000. However, the United States is paying about 35 percent *more* for that ship in constant 2014 U.S. dollars. The ships in FY 2000 were part of a three-ship buy, whereas FY 2014 was a one-ship buy, making the FY 2000 ships much more affordable. The answer using the November exchange rate to buy the ship in 2014 is about 7 percent less expensive. In four months, the exchange rate changed such that the difference was cut in half. The point is that fluctuations in exchange rate could partially reduce or magnify any domestic shipbuilding premium in Australia and add additional budgetary risk.

Table 5.12
Australian Costs to Purchase a DDG-51-Class Destroyer in 2000 and 2014

Fiscal Year	Country Purchasing	Unit price (USD millions)	Exchange Rate to Currency[a]	Cost to Australia (AUD millions)	Escalation Rate to 2014	Cost in Final Currency (2014 $)
2000	Australia	925	1.72[b]	1,573	1.51	AUD 2,379
2014 (July)	Australia	1,926	1.06[c]	2,042	1.00	AUD 2,042
2014 (November)	Australia	1,926	1.15[c]	2,215	1.00	AUD 2,215
2000	United States	925	1.0	N/A	1.55	USD 1,433

[a] Midpoint exchange rate data from OANDA Corporation, "Historical Exchange Rates," undated.
[b] Annual average.
[c] Monthly average.

A sense of recent exchange rate differences between Australia and other currencies is seen in Figure 5.3, which shows the exchange rates of Australian dollar to U.S. dollar and Australian dollar to euro for the past five years (by month) based on data from OANDA Corporation. The volatility of the rates (represented by a standard deviation) is about 6.5 percent for the U.S. dollar and 5.2 percent for the euro. Such differences are significant when compared with any shipbuilding premium to build in Australia on the order of 40 percent.

Summary of Cost Benchmarking

A variety of different benchmarks and methods indicate that Australian naval shipbuilding tends to be more expensive than our comparator countries: Italy, Japan, Korea, Spain, the United Kingdom, and the United States. The results are remarkably consistent despite the limitations and caveats concerning the data used. Nonetheless, one should not overly interpret the precision of the values. In Table 5.13, we summarize the premium for the different metrics relative to a U.S. basis. The range is roughly 20 percent to 45 percent higher relative to the United States. This range seems to depend on ship type (although there are not enough observations to be definitive). The combatants (frigates and destroyers) seem to have a consistent premium of around 30 percent to 40 percent. The amphibious ship premium is lower, at about

Figure 5.3
Exchange Rates for Australian Dollar to U.S. Dollar and Euro over the Past Five Years

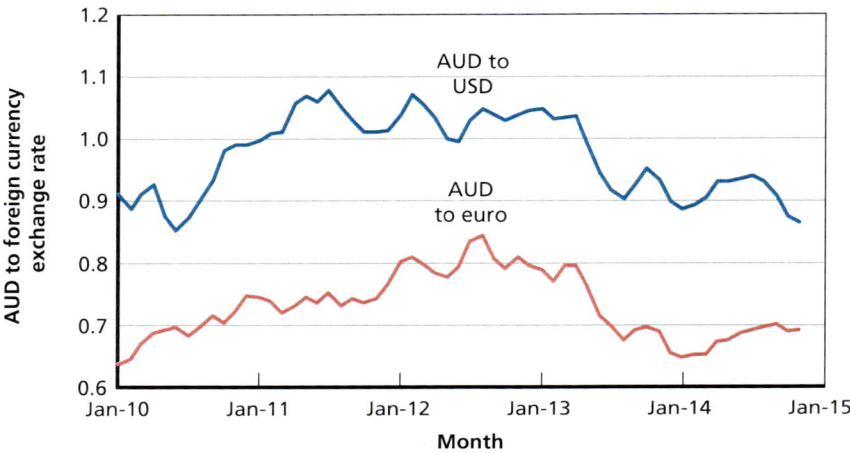

SOURCE: OANDA Corporation, undated.
RAND RR1093-5.3

12 percent more than a U.S. basis. Again, the CPT metric is less robust for amphibious ships and reflects that a significant portion of the ship has been built in Spain.

Overall, the various methods all indicate a modal Australian naval shipbuilding premium of about 30 to 40 percent for ships built entirely in Australia. Another important consideration is that any consideration of foreign or domestic build must consider the exchange rate risk, which can significantly influence the perceived premium to build in Australia.

Schedule Benchmarking

Approach

Warships are extremely complex weapon systems. Their design and construction takes time, and in several ways, they are different than other weapon systems. Unlike other weapon systems, there are no test articles or prototypes; every ship that is constructed during the design

Table 5.13
Summary Metrics for Australian Shipbuilding Costs Relative to a U.S. Basis

Method	Metric	Approximate Australian Premium Relative to a U.S. Basis (%)
Input	Direct shipbuilding labor wages	40
	Manufacturing labor costs	35
	Oil and gas industry construction	20
	Construction cost adjusted to First Marine International shipbuilding productivity[a]	45
Comparative	Frigate costs	40
	Destroyer costs	30[b]
	Amphibious ship costs	12[c]
Parametric		35

[a] Cost comparison based on hours per compensated gross tonnage, a productivity measure used by First Marine International Ltd. This measure compares the weights of different types of commercial ships with one another by using adjustment factors that depend on the ship type (e.g., tanker, dry cargo, ferry). See First Marine International Ltd., *First Marine International Findings for the Global Shipbuilding Industrial Base Benchmarking Study*, Part 2: *Mid-Tier Shipyards, Final Redacted Report*, February 6, 2007.

[b] Prior to rebaseline.

[c] Based on the recent LHD. Because significant portions of the ship are built in Spain, the relative costs may not be representative of a complete Australian build (the premium is likely lower than if the ship had been fully built in Australia).

and build phase will enter service. In fact, the fabrication and assembly of warships begins soon after the start of the design phase. Often, formal acquisition processes for ships are highly tailored because warships have relatively lengthy design-to-build schedules.[22]

To compare schedule across various warships, we examine the length of time between the following key milestones:

- contract award date, when the contract is awarded for design and construction

---

[22] See Jeffrey Drezner, Mark V. Arena, Megan P. McKernan, Robert E. Murphy, and Jessie Riposo, *Are Ships Different? Policies and Procedures for the Acquisition of Ship Programs*, Santa Monica, Calif.: RAND Corporation, MG-991-OSD/NAVY, 2011.

- keel date, when the keel is laid down, which is the start of major assembly of the ship (the actual construction of the parts begins earlier in the development schedule; we treat this date as a proxy for the start of major construction of the ship)
- delivery date, when the ship is delivered to the client
- commissioning date, when the ship enters the service.

The schedule data for which we were able to find consistent information from the U.S., Australian, and European ships were for the keel date, or the start of major construction and assembly. The other schedule point of comparison is the date when the ship was commissioned. The European ships included in the analysis are French and Italian FREMMs, the Royal Dutch Navy's LCF, and Britain's Type 45 destroyer (T-45). The U.S. ships included are the *San Antonio* class (LPD-17), U.S. DDG-51 class, *Wasp* class (LHD-1), amphibious assault ship, and *Oliver Hazard Perry*–class frigate. We also included some limited data from the Japanese *Akizuki* class, as well as the Korean *Incheon* class and *Sejong Daewang* class. In addition, we included Australian AWDs and the *Anzac*-class frigates.

Caveats

This analysis is largely based on open-source data. As much as possible, we have tried to validate the information using several sources. In addition, we were unable to meet with the contractors or government program offices responsible for the design and manufacturing of the ships; therefore, we assumed that each contractor defines each of the major schedule milestone in the same way. The data include a combination of various kinds of warships. There are more frigates and destroyers than amphibious assault ships. We remind the reader that much of this analysis is for comparative and benchmarking purposes only.

Source of Data and Initial Analysis

Data used in this analysis are drawn from open sources, such as the Naval Vessel Register (NVR),[23] *Jane's Fighting Ships*,[24] and other open

---

[23] Naval Vessel Register, "Ships," web tool, undated.

[24] IHS, undated.

sources available on the Internet for the U.S. and European examples. The Australian ship schedule data come directly from data provided by the White Paper team, augmented by information available online via the *Jane's Fighting Ships* website.

Table 5.14 compares the number of months from keel to commissioning for each of the ships.

Table 5.15 compares the U.S.-contracted ship's keel-to-commission schedule. As one can observe, the number of months is as high as about

Table 5.14
Number of Months from Keel to Commissioning Table of Means

Program	Minimum	Mean	Maximum	Number of Observations
T-45	64	75	80	6
F590 FREMM	46	56	63	8
D650 FREMM	52	67	84	8
De Zeven Provinciën LCF	42	43	45	4
LPD-17	51	60	68	9
LCS	37	44	50	4
DDG-51	27	33	49	62
FFG-7	17	27	41	51
Iver Huitfeldt	44	53	58	3
Anzac	31	43	54	8
Akizuki	33	36	45	4
Incheon	49	49	49	1
Sejong Daewang	50	44	38	2
LHD-1	41	48	69	8

Table 5.15
Keel-to-Commission Schedule for U.S. Warships (months)

Program	Minimum	Mean	Maximum	Number of Observations
LPD-17	51	60	68	9
LCS	37	44	50	4
DDG-51	27	33	49	62
FFG-7	17	27	41	51
LHD-1	41	48	69	8

70 months to as low as about 20 months. DDG-51 and FFG-7 classes are interestingly well below the other average values in the table. This is probably due to their long production history, higher rate of production (two to four ships per year), and greater total quantities produced.

Figure 5.4 compares the months from start of construction or keel to commission of each ship. As one can observe, the ship months of construction for most of the ships are not reduced with time as one might expect from a traditional production, where manufacturing and assembly time is reduced as production workers learn and become more efficient and as improvements in manufacturing and assembly processes are introduced.[25] However, when design changes are introduced during production, and if these changes are significant enough, they negate the effect of the learning, which has occurred through the construction of the earlier ship. Figure 5.4 may be capturing the impacts of design changes and performance upgrades during each ship construction, which occur from ship to ship. With the exception of

Figure 5.4
Time-Series Plot of Keel-to-Commission Schedules

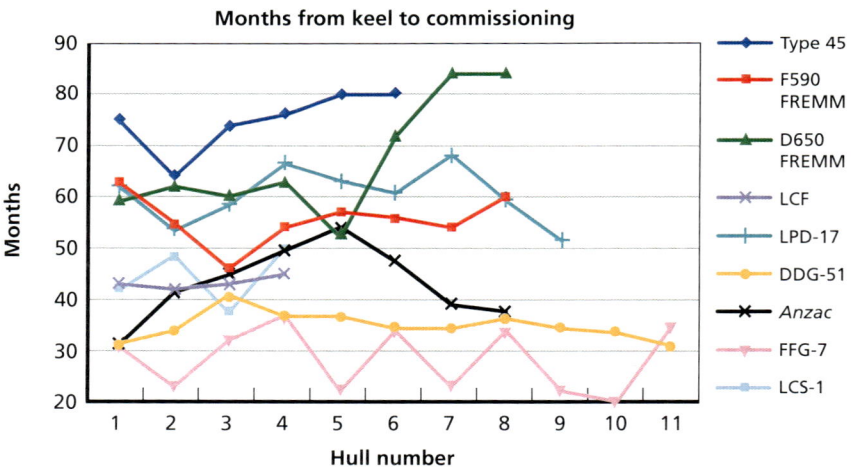

RAND RR1093-5.4

---

[25] This phenomenon is often described as learning by doing. The phenomenon occurs when the systems being produced are relatively similar to each other.

DDG-51, which shows a steady decline in time after the third hull, the remaining ships' keel-to-commission times unevenly increase or decrease. For keel-to-commission times available from the Naval Vessel Register, the DDG-51 and FFG-7 data go beyond Hull 11, but for this illustrative and comparative purpose, we truncated the data.

In the next plot, we compare the schedule of two Australian ships—the *Anzac*-class frigate and the AWD—with the European and American ships discussed earlier. Figure 5.5 shows the average keel to commissioning in months. The Australian AWD is on a slightly different basis—start of fabrication to the delivery date. However, these dates should be within one or two months of keel to commissioning and therefore comparable with the rest of the schedule data. Another caveat on the AWD data is that they are projections and not actual deliveries. The average keel to commissioning for the *Anzac* class was 42 months, which is well below the average of about 48.5 months, and the AWD is projected to be slightly above the overall average.

Figure 5.5
Average Keel-to-Commission Schedule

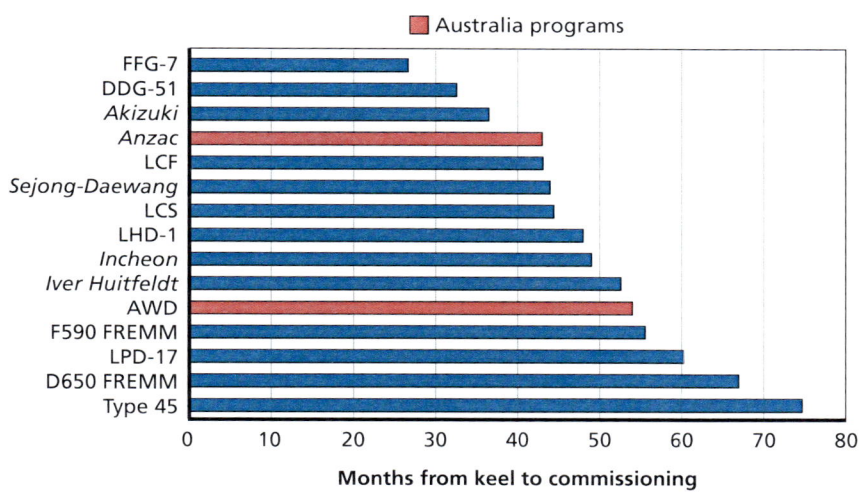

Weight as a Proxy for Complexity

Weight is often used as a proxy for complexity of a weapon system. In Figure 5.6, for each ship, we present a plot of the first ship's full load displacement versus its months from keel to commissioning. We used the scheduled keel to commissioning for the first ship to eliminate the effects of learning and the quantities of ship produced. We would highlight that the quantities of DDG-51 and FFG-7 produced were substantially higher than any of the other ships included in our data. In addition, the full displacement load of LPD-17 and LHD-1 are much higher than the rest of the ships included in the analysis. The plot does not show a meaningful trend.

Summary of Schedule Benchmarking

We compared the number of months it took from keel to commissioning for a variety of European and U.S. ships with two Australian ship classes, as well as with a ship from both Japan and Korea. The average

Figure 5.6
First Ship's Metric Tons of Full Load Displacement Versus Months from Keel to Commissioning

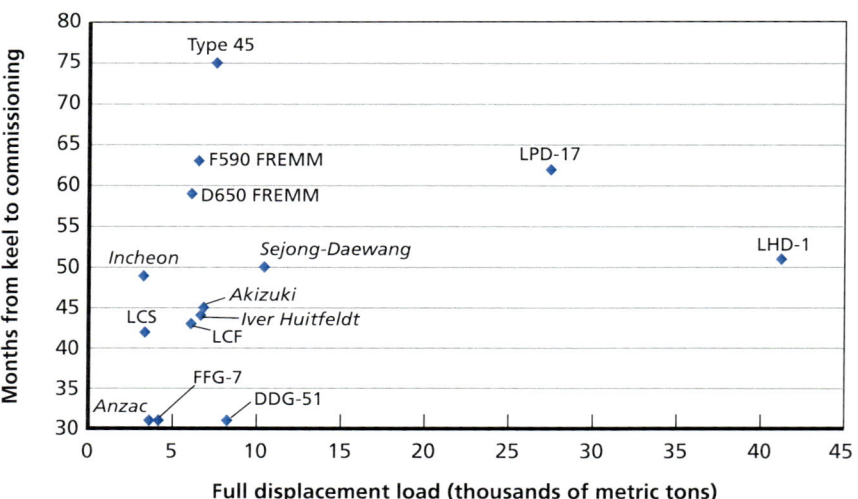

time of the Australian *Anzac* class is faster than the average of the other ships that we included in the analysis. AWD average time is comparable with the average of all the ships that we analyzed.

Observations on Australian Shipbuilding Cost and Schedule

Overall, the cost and schedule benchmarking results suggest that Australian shipbuilding is significantly more expensive with respect to cost and slightly longer with respect to schedule. Could Australian shipbuilding do better? We believe the answer is yes, cost performance could be improved. The input cost benchmarking analysis suggests what may be possible for Australian shipbuilding. Based on the Compass International data of oil, chemical, and gas construction industry costs, our analysis results indicate around a 20 percent premium—two-thirds to one-half of the shipbuilding cost premium. Because this industry is similar to shipbuilding (and uses many analogous trades), such a premium should be a reasonable target for the Australian shipbuilding industry to achieve. It would be unreasonable to expect world-class labor productivity, and not much can be done to influence wage rates and material costs. Much of the improvement would be geared toward better productivity. Once productivity improves, schedules are likely to be more competitive as well.

As we saw in the previous chapters, a sustained build program would help to develop and retain skilled workers—and thus improve productivity. But the needed improvements go beyond just more-proficient workers; many acquisition practices also have to improve. One necessary change is a much more rigorous approach to program execution to avoid the issues seen on the AWD program.[26] These improvements include better integration between designers, builders, and suppliers; a mature design at the start of the build; and control of requirements and design changes once building begins.

---

[26] Australian National Audit Office, 2013.

Another important change to enable this productivity improvement is to engage in a continuous build philosophy for the shipbuilding program. As was described in earlier chapters, such an approach avoids a "boom-bust" cycle for industry and allows industry to maintain and train a skilled workforce. Continuity of work also allows the shipbuilders to justify investments to achieve better productivity, because there is a dependable, long-term cash flow. On industry's part, there are some changes that are enabled by a continuous build philosophy. One such change is a cultural shift to a continuous improvement philosophy. Such a change was seen in the United States submarine-building industry at the end of the Cold War.[27] The industry recognized that its products were becoming unaffordable and made radical changes to the way it designed and built submarines, focusing on cost-effectiveness. Very strong and visionary leadership at the companies drove this change from the top.

However, getting this better performance would not happen overnight and might take several years to develop. Making all the changes described above could reduce the cost premium during the Future Frigate program. It would not happen initially, but we speculate that the premium could be halved by the third or fourth hull.

---

[27] See John F. Schank, Cesse Ip, Francis W. LaCroix, Robert E. Murphy, Mark V. Arena, Kristy N. Kamarck, and Gordon T. Lee, *Learning from Experience*, Vol. II: *Lessons from the U.S. Navy's* Ohio, Seawolf, *and* Virginia *Submarine Programs*, Santa Monica, Calif.: RAND Corporation, MG-1128/2-Navy, 2011.

CHAPTER SIX

Examining Economic Pros and Cons of Australian Government Investments in Various Naval Shipbuilding Enterprise Options

This chapter uses analogies from the United States and Sweden to shed light by implication on the economic consequences of shipbuilding in Australia. It is built around three case studies that were informed by the extensive literature on economic multipliers: Newport News Shipbuilding in Newport News, Virginia; Austal USA in Mobile, Alabama; and the Gripen program undertaken by Saab Aeronautics in Linkoping, Sweden.

A full discussion of the literature and case studies is contained in a companion RAND report.[1]

Economic Multipliers and Their Uncertain Implications

While there is a sizable literature on economic multipliers, its implications for the economic consequences of shipbuilding in Australia are uncertain. The basic logic of an economic multiplier is straightforward. Suppose that the government spends $100 buying a good or service from a shipyard. The shipyard might then be expected to spend at least a portion of that money on inputs, such as labor or materials. The original $100 creates a cascade (i.e., multiples) of spending through the economy; that is, $100 spent at a shipyard results in additional spend-

---

[1] Edward G. Keating, Irina Danescu, Dan Jenkins, James Black, Robert Murphy, Deborah Peetz, and Sarah H. Bana, *The Economic Consequences of Investing in Shipbuilding: Case Studies in the United States and Sweden*, Santa Monica, Calif.: RAND Corporation, RR-1036-AUS, 2015.

ing by shipyard workers at local restaurants, which then hire additional workers who rent additional housing, and so forth.

Several studies have estimated economic multipliers associated with defense spending. Most of the resulting estimates are in the range of 1.7–1.9—that is, $100 spent at a shipyard ultimately results in $170–$190 worth of additional economic activity in the shipyard's region (inclusive of the original $100).

Economic multipliers may be lower (i.e., less than 1.0) if the increased spending displaces other economic activity. Studies looking at World War II often find multipliers less than 1.0 because increased defense spending displaced private-sector spending. On the other hand, if the spending at the shipyard results in favorable spillover effects into the economy (e.g., spin-offs into other industries), one could find an economic multiplier greater than the 1.7–1.9 range.

Newport News Shipbuilding Case Study

Newport News Shipbuilding (NNS) is the largest private-sector single-site employer in the Commonwealth of Virginia and a major economic engine of the Hampton Roads region. RAND's examination utilized extensive subject-matter expert interviews, open literature, and publicly available data.

NNS is an "employer of choice" in its region. NNS pays its employees well, with only limited annual attrition. Shipyard jobs tend to be considerably more desirable than most workers' next-best alternative, especially in light of most workers' reluctance to geographically relocate.

NNS appears to have generated relatively few local spin-offs. Experts are concerned that the Hampton Roads region, in general, lacks a heritage of entrepreneurial behavior.

The area immediately proximate to NNS is not economically vibrant. Experts told us that NNS workers rush to their automobiles at the end of their daily shifts (3:30 p.m.) and leave the immediate vicinity as quickly as possible.

Austal USA Shipbuilding Case Study

Whereas NNS is a long-established shipyard, Austal's operations in Mobile, Alabama, only developed in earnest in the past ten years. Austal's scale of operation increased by nearly a factor of five between 2009 and 2014 (though it remains considerably smaller than NNS, in terms of both revenue and employment level).

Most Austal employees live in Mobile County or nearby Baldwin County, Alabama, but a sizable portion commutes from the neighboring states of Mississippi and Florida. Reflecting the fact that shipyard jobs are both unique and relatively well paying, we have consistently found a willingness on the part of shipyard employees to undertake sizable driving commutes.

In order to obtain required training for its growing workforce, Austal USA has relied on the Maritime Training Center, funded by the state of Alabama. Between individuals who are trained at the Maritime Training Center but not ultimately hired by Austal and considerable attrition at Austal USA, the company has had the effect of sizably altering the workforce skill profile in the greater Mobile area beyond its current employees. Austal has not (at least yet) caused development of a network of proximate local suppliers.

Gripen Case Study

Sweden's JAS-39 Gripen fighter program has been lauded for successfully delivering an advanced fighter aircraft while also producing a significant economic multiplier to the local and national economy. It has been extensively cited in discussions of Australia's shipbuilding industry.[2] The RAND research team therefore conducted a literature review and subject-matter expert interviews to examine the Gripen program's wider benefit to Sweden.

---

[2] See, for example, Goran Roos, "Future of Australia's Naval Shipbuilding Industry," supplementary submission to the Senate Economics References Committee, October 13, 2014; and Economic Development Board South Australia "Economic Analysis of Australia's Future Submarine Program," October 2014.

Commenced in the early 1980s, the Gripen aircraft was produced by Saab in Linkoping in central Sweden, about 170 km southwest of Stockholm. The program originally had a target for creating 800 jobs in a region with high unemployment. By 1987, the program had generated an estimated 1,200 new jobs. Today, the program is thought to sustain roughly 3,000 jobs in Sweden, with hopes to market a "next generation" upgrade of the Gripen through to 2040. Anchored around Saab, the local technical university, and a number of science parks, the wider Linkoping "aerospace cluster" currently employs around 18,000 workers, which is approximately one-third of the local workforce. Many academics have argued that the program has generated significant knowledge spillovers and a variety of spin-off firms, several working in areas quite distant from aviation.[3] The program is also credited with helping to sustain established firms, such as Volvo and Ericsson. The Gripen program appears to have had a larger (more favorable) economic multiplier, estimated by Eliasson (2010) to be around 3.6, than the 1.7–1.9 range more typically found for major defense projects.

Discussion

It is impossible, lacking greater specificity, to estimate the economic consequences of a shipbuilding project on a region of Australia or on the nation as a whole. Rather, the applicable economic multiplier is a highly contextually dependent question. If shipyard work displaces skilled workers from other high-value activities, the economic multiplier could be less than 1.0.

However, our examination of shipyards in the United States suggests an economic multiplier greater than 1.0. At the shipyards we examined, expert interviews suggested that the shipyards seem able to attract job applicants, suggesting that these workers do not have alter-

---

[3] The most notable academic is Gunnar Eliasson. See Gunnar Eliasson, *Advanced Public Procurement as Industrial Policy*, New York: Springer, 2010; and Gunnar Eliasson, "The Commercialising of Spillovers: A Case Study of Swedish Aircraft Industry," in Andreas Pyka, Derengowski Fonseca, and Maria da Graca (eds.), *Catching Up, Spillovers and Innovation Networks in a Schumpeterian Perspective*, New York: Springer, 2011.

native employment options as desirable as working at the shipyards. The shipyards have not displaced high-value activities for these workers, consistent with a larger economic multiplier from shipbuilding.

On the other hand, the high level of spillovers and spin-offs seen in the Gripen case study are not consistent with experiences at U.S. shipyards. NNS appears to have generated relatively few spillovers. Indeed, the entire Hampton Roads region has been critiqued for a dearth of entrepreneurial activity. Likewise, no cluster of suppliers has yet emerged around Austal USA. Evidence from U.S. shipbuilding suggests that the Gripen example is overly optimistic from an economic development perspective. But the fact that neither shipyard has had Gripen-like effects should not obfuscate the fact that both have provided considerable favorable economic consequences to their regions.

CHAPTER SEVEN

Conclusions and Recommendations

Over the next 20 years, as we have noted in the previous chapters, AUS DoD intends to acquire upward of 50 naval surface ships and submarines. These acquisitions will include 13 to 15 large surface ships, such as AWDs, LHDs, and Future Frigates, as well as 27 to 35 smaller ships, such as patrol boats, OPVs, and LMRVs.[1]

Successfully realizing these acquisitions is likely to require the Australian government to commit to a fundamental industrial policy decision when it releases the 2015 Defence White Paper. The government must choose to build the naval surface ships on Australia's acquisition list entirely in-country, build them partially in-country and partially overseas, or have them built at shipyards overseas. Each strategy carries costs and risks; none is wholly advantageous or can be put into effect overnight.

Detailed Findings

In the remainder of this chapter, we summarize in greater detail our answers to the four motivating questions that were posed to us by the White Paper Enterprise Management team and that we discussed in Chapter One.

---

[1] For the purpose of this analysis, the distinction between patrol boats, OPVs, and LMRVs was used for modeling purposes only. Australia's Force Structure Review process will consider the requirements to address these smaller vessels.

What Are the Comparative Costs Associated with Alternative Shipbuilding Paths?

We examined three potential paths that the Commonwealth might take for the indigenous shipbuilding industrial base.

- Path 1, used for *Hobart*-class destroyers, is to build ships in Australia using a fully capable domestic shipbuilding industry.
- Path 2, used for the *Canberra*-class LHDs, is to build major portions of a new ship, such as HM&E equipment, in another country, with Australian shipyards completing construction and installing major weapon and combat systems.
- Path 3, pursued in the acquisition of the *Oberon* class of submarines, is to build and outfit an entire RAN warship class overseas.

Path 1

If Australia were to choose Path 1, it faces a large gap between the production of the last AWD and the first Future Frigate. We depicted that gap in Figure 4.2, assuming the Future Frigate starts construction in 2020. To overcome that gap with a wholly domestic industry, Australia can alter the number of ships it would build domestically, their start dates, and build durations compared with the current shipbuilding plan (which consists of three AWDs and eight Future Frigates). Several of those variants are portrayed in the first column of Table 4.3, which outlines our estimate of their total shipyard labor costs, including overhead, training and termination, and schedule delay in replacing retiring *Anzac*-class ships.

There are several take-aways from Table 4.3. The costs of the various options for lessening the gap do not vary much from the base case costs, and the differences fall within our estimating error. The bigger difference is in the total delay in replacing *Anzac*-class ships as they retire. Large numbers of unproductive workers must be hired in the base case to meet the Future Frigate demands. The unproductive labor causes schedule delays in delivering the frigates (see Figure 4.6) when nothing is done to lessen the gap. Starting the Future Frigate construction early helps address this problem, although a construction start prior to 2020 is highly unlikely. Starting a fourth AWD almost immediately also greatly reduces the schedule delay, but there is no requirement for a fourth AWD and the non-labor cost of the ship is very high. The most promising option for lessening the gap is to build some number of OPVs in the gap. If four OPVs are built starting in 2017, costs increase by AUD 120 million over the base case of doing nothing, but four additional OPVs are added to the RAN fleet at very marginal costs. Furthermore, delays in delivering replacement ships are reduced to almost zero.

Figure 7.1 displays these relatively similar costs, while Figure 7.2 shows the schedule delays (in both figures, the various options that we investigated are juxtaposed next to Australia's current baseline plan, portrayed in the red bars, of producing eight Future Frigates beginning in 2020 as outlined in the *2013 Defence White Paper*). Figure 7.1 shows that all strategies would cost some AUD 5.5 billion, strongly suggest-

**Figure 7.1
Total Labor Costs of Base Case and Alternative Shipbuilding Construction Paths**

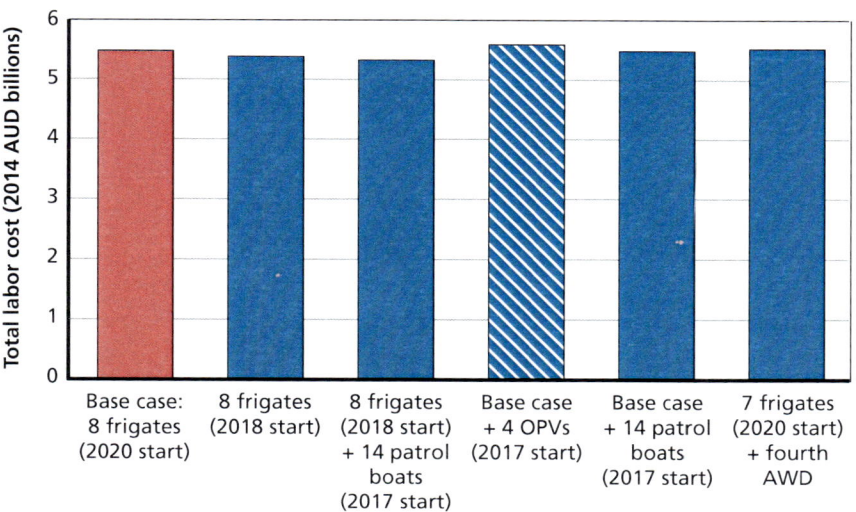

NOTE: The figure assumes that the base case Future Frigate uses 5 million man-hours, takes 6.5 years to build, and has a 95-percent unit learning curve; in addition, the last six ships are produced with a drumbeat of one.
RAND RR1093-7.1

ing that keeping as much as 30 percent of shipyards' Future Frigate workforces employed during the gap years would not be much costlier than allowing worker headcounts to drop to zero. In addition, it shows that lessening the gap by building OPVs (portrayed in the crosshatched bar) would provide additional ships to RAN at a very marginal labor cost to produce them. Figure 7.2 shows that most options for lessening the gap would significantly reduce the total delay in delivering *Anzac*-class replacements.

If nothing is done in the gap between the end of the AWD construction program and the start of Future Frigate construction in 2020, it will be difficult to sustain more than a single shipbuilder in Australia. Lessening the gap with OPVs, for example, will help in sustaining more than one shipbuilder if each shipbuilder is given an OPV to build every year (a total of six to eight OPVs). And the Future Frigate pro-

Figure 7.2
Total Schedule Delay of Base Case and Alternative Shipbuilding Construction Paths

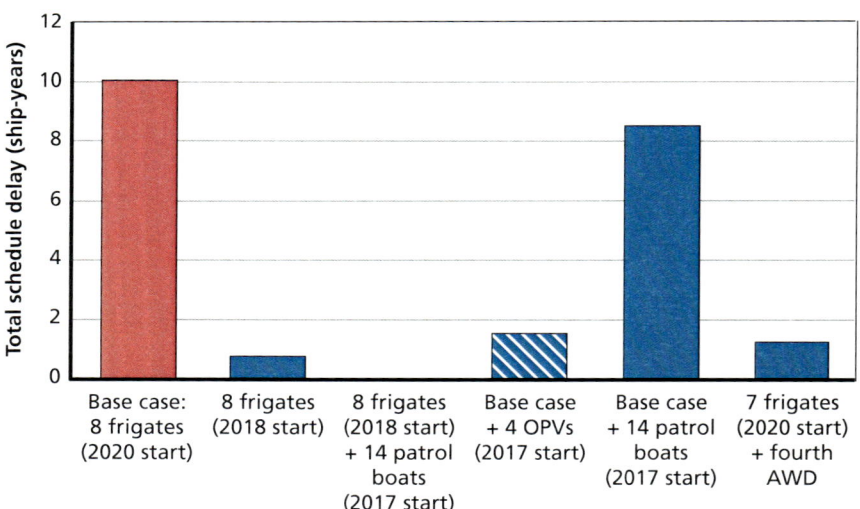

NOTE: The figure assumes that the base case Future Frigate uses 5 million man-hours, takes 6.5 years to build, and has a 95-percent unit learning curve; in addition, the last six ships are produced with a drumbeat of one.
RAND RR1093-7.2

gram should have sufficient demand to sustain two shipbuilders—one that builds blocks and one that both builds blocks and assembles them into a completed ship. Given the low demand for new RAN ships, it will be difficult and expensive to sustain more than two shipbuilders.

Path 2

As Figure 4.25 shows, a similar gap in demand exists for skilled workers in Australian shipyards that are involved in completing construction and installing major weapon and combat systems on partially built ships obtained from overseas.

We examined the same options for closing this gap as for closing the Path 1 gap. Table 4.6 shows the same shipyard labor component cost comparisons for the options that we showed in Table 4.3. It suggests that starting the Future Frigate earlier reduces the total cost by a small margin because the efficiency of the workforce would improve on a cost-per-FTE basis. In general, adding a fourth AWD increases total cost, as would be expected given the bigger ship.[2]

Adding patrol boats increases the efficiency on a cost-per-FTE basis, but the total cost is still greater than the baseline, as would be expected given that more ships are being produced. Also, building patrol boats or OPVs sustains mostly structural skills that are not utilized to a high degree in Path 2. Path 2 involves primarily the final outfitting of the Future Frigates, and building patrol boats or OPVs in the gap does little to sustain those skills.

For Path 2, two shipyards could be sustained if new ships are delivered every year. If the drumbeat expands to a new ship every two years (discussed below in the section on a continuous build strategy), it would be difficult to sustain more than one shipbuilder.

Path 3

The third path open for the future Australian shipbuilding industrial base is to basically abandon new ship construction in Australia and buy as-built ships from other nations. Some specific system work may be accomplished in Australia, but the Australian ship-related resources

---

[2] We assume that the fourth AWD is a full build in Australia, unlike the Future Frigates, which are final construction and outfitting only.

Table 5.5
Comparison of Physical Characteristics and Hulls Produced, Amphibious Vessels

Ship	Country	Length (m)	Beam (m)	Draft (m)	Full Load Displacement (metric tons)	Number of Hulls Produced
JDS *Izumo* (DDH)	Japan	248.0	38.0	7.3	24,000	0[a]
Canberra LHD	Australia	230.8	32.0	7.1	27,500	2
Juan Carlos (L61)	Spain	230.8	32.0	6.9	26,000	1
San Antonio-class LPD-17	United States	208.5	32.0	7.0	25,883	10[b]
America-class landing helicopter assault (LHA)-R	United States	257.3	32.3	8.5	44,449	1[c]
Wasp-class LHD-1	United States	258.2	36.0	8.5	41,684	8
Albion-class LPD-R	U.K.	176.0	28.9	7.1	19,560	2

SOURCE: IHS, undated.
[a] Two in production.
[b] Completed, one hull in construction.
[c] Improved design of one previous hull.

For each ship, we calculate a relative cost index by type. We summarize the index calculation steps as follows:

1. Calculate an average unit procurement cost (i.e., removing nonrecurring engineering costs), as necessary.[17] The U.S. ships have much longer production histories. For these ships, we use a mix of data points. For the FFG-7 class, we use a total program average (low) or an average of the last ten ships (high). The last ten are more expensive because they include both additional upgrades that occurred over production and significant end-of-production costs. For the DDG-51 class, we use fiscal year (FY) 2010 and FY 2011 authorization values, because these points

---

[17] It would be more accurate to calculate cost normalizing for cost improvement—that is, to a common point along the improvement curve. However, much of the cost data are for the total program and not at an individual hull level. Thus, we use average unit procurement instead.

and capabilities would be concentrated on supporting in-service ships rather than building new ones. As described in Chapter Two, sustaining a ship support industrial base is a function of the naval fleet and the policies for maintaining that fleet. Our initial analysis suggests that the current and future plans should adequately sustain an in-service ship support industrial base.

A Potential Fourth Path

By building and outfitting first-of-class vessels in overseas shipyards and subsequent vessels in Australia, this path seeks to mitigate design and production problems that first vessels often encounter. This path has been used to some extent with Australia's *Adelaide* and *Huon* programs. By following this path, the *Adelaide* program obtained a highly mature design from the original, experienced producer in the United States (Todd Shipyards in Seattle, Washington, now known as Vigor Shipyards). The ship's designer, the U.S. firm Gibbs & Cox, kept the technical information current as the U.S. design evolved. During initial production, Australia stationed key personnel in the United States who had constant access to the design and to Todd's practices. When production moved to Australia, Todd kitted virtually all material (down to fasteners, such as nuts and bolts), which it shipped to Australia. It also provided detailed assembly (production) documentation and seconded key personnel. So, too, did the U.S. supplier of the combat system, which provided integration and testing services on site. In so doing, this approach was able to use a mature design, proven production sequences and procedures, material that was kitted and provided to the construction yard, and readily available on-site construction and testing services when production began in Australia.

While attractive, this path would increase gaps in demand for certain shipbuilding industry workforce skills compared with Path 1, particularly if the overseas first-of-class production schedule were to be lengthy or protracted. Moreover, it would entail additional costs for Australia to customize the original design used overseas into production instructions that can be used in Australian shipyards.

Is It Possible for Australia's Naval Shipbuilding Industrial Base to Achieve a Continuous Build Strategy, and How Would Such a Strategy's Costs Compare with the Current and Alternative Shipbuilding Paths?

In shipbuilding, drumbeat refers to how frequently new ships are delivered. For example, a drumbeat of one implies that a new ship is delivered every 12 months. In the short term (2015 to 2030), the drumbeats are determined by the need to replace ships currently in the RAN force structure. For example, the last six *Anzac*-class frigates were commissioned at the rate of one per year, suggesting that the new frigates that will replace the *Anzac* class will be needed at the same rate (i.e., a drumbeat of one). The major ship force structure resulting from different drumbeats and average ship lives is shown in Table 4.4.

Given that Australia's currently planned naval force structure comprises 14 to 16 major surface ships (including three AWDs, eight to ten Future Frigates, two LHDs, and an LSD) and 27 to 35 smaller vessels (patrol boats, OPVs, and LMRVs), drumbeats that deliver ships every 24 months or longer will probably not sustain a desired future force structure of a fleet whose average life is 30 years. Deliveries every 30 months can work if Australia were to adopt ship lives of 35 or 40 years. Shorter ship lives would require more-frequent construction starts (i.e., drumbeats of less than two).

Our analysis suggests that Australian domestic naval shipbuilders can sustain an 18- to 24-month pace of large ship construction starts if AUS DoD carefully manages Future Frigate deliveries and keeps those ships operational for 25 to 30 years. With regard to the smaller vessels in the RAN fleet, Australian industry would have to maintain a pace of one or more construction starts a year if RAN were to assume that those vessels would be operational for up to 30 years.

How Do the Costs of Acquiring Vessels Domestically Compare with Acquiring Comparator(s) from Shipbuilders Overseas?

Numerous benchmarks and methods indicate that Australian naval shipbuilding tends to be more expensive than the comparator countries we used in this analysis: Italy, Japan, Korea, Spain, the United Kingdom, and the United States. Table 5.15 summarizes the premium relative to a U.S. basis for different metrics that we examined. The premium ranges between 20 percent and 45 percent, and the range seems to depend on ship type (although there are not enough observations to be definitive). Combatants seem to have a consistent premium of around 30 percent to 40 percent. The amphibious ship premium is lower, at about 12 percent more than a U.S. basis. The CPT metric is less robust for amphibious ships and reflects that a significant portion of the ship has been built in Spain.

Another important consideration is that any decision on foreign or domestic build should take exchange rate risk into account, because this can significantly influence the perceived premium to build in Australia.

With regard to schedule benchmarking, we compared the number of months it took from keel to commissioning for a variety of European and U.S. ships with two Australian ship classes, as well as with ships from Japan and Korea. The average time of the Australian *Anzac* class is faster than the average of the ships that we included in the analysis. AWD average time is comparable to the average of all the ships that we analyzed.

We believe that this shipbuilding premium could be cut in half if the following changes are made:

- Engage in a continuous-build strategy.
- Improve acquisition practices to have more-mature designs at the start of construction and to minimize change during construction.
- Encourage industry to shift to a continuous-improvement culture.

Achieving this better performance would not happen overnight and might take several years to develop. However, we feel that it is achievable mid-way through the build of the Future Frigate program.

How Much Do Expenditures Connected with Warship Building, Maintenance, and Sustainment Add to Australia's Economy?

Australia can either purchase its warships internationally or produce them domestically. As shown, there is a price premium associated with the latter approach, which would imply greater tax burden associated with indigenous production. Note, however, that conditional on having decided to purchase a warship, there will be societal costs of taxation under either approach. Those taxation costs will be greater, however, if a decision is made to pay a price premium for indigenous production.

From an economic perspective, there may be advantages to indigenous production that could offset increased taxation costs. We put these prospective advantages into two categories that we label *favorable spillovers* and *increased workforce utilization*.

Favorable spillovers would occur if the process of shipbuilding gives rise to ancillary benefits. In the Gripen case in Sweden, the program is credited with leading to technological developments in many realms, some quite far removed from shipbuilding (including, indeed, a firm that produces advanced solutions for tooth implants). Sweden's Gripen program is credited with energizing the so-called Linkoping *aerospace cluster* that has clearly had a transformational effect on its region.

Unfortunately, RAND's analysis of shipbuilding in the United States did not find favorable spillovers in the fashion of Gripen. Shipbuilding has been favorable to local economies, but it has done so in a more modest fashion, without the ecosystem of favorable spin-offs and spillovers associated with Gripen. We do not think an outcome from shipbuilding similar to that in Silicon Valley from technology is a realistic aspiration. The Gripen analogy is inapt.

Where U.S. shipbuilding has had favorable impact is in the realm of increased workforce utilization. In particular, the individuals who are employed at shipbuilders in the United States in many cases would otherwise be in much lower-paying, lower-skilled jobs, or might not be in the paid labor force at all absent the shipbuilder. U.S. shipbuilders indicated that, when they recalled workers who had been laid off, many of those workers readily returned. Shipbuilding provides economic opportunities to individuals whose skills might not otherwise be commensurably well-employed in the U.S. economy.

The extent to which this increased workforce utilization argument applies to Australia is highly context-dependent. If the Australian economy or, more specifically, the economy in the area where the ships would be built is already at full employment, workers hired by the shipyard would simply be displaced from other gainful employment. There would be no increase in workforce utilization. Rather, workers would simply be reallocated from other useful pursuits to no net societal advantage.

But full employment does not currently characterize the regions we studied proximate to shipbuilders in the United States, and, to the extent prospective Australian shipbuilding regions are similar, there may likewise be increased workforce utilization advantages associated with indigenous shipbuilding in Australia.

Summary Implications

While Australia has a history of indigenous naval shipbuilding that dates back more than 100 years, it has swung between acquiring naval ships from domestic shipbuilders and obtaining them from producers abroad, and it has had several "boom-bust" cycles. Having a robust shipbuilding industry offers numerous advantages. It supports local industry, gives the Australians control over design and production, and creates a skilled labor pool that can spill over into other venues. But the demand for naval ships is limited, and thus demand may not be robust enough to support a full-fledged industry. One result of relatively low demand is a shallow pool of expertise in terms of design and production and a less experienced workforce than would be available in a global market. Also, maintaining a local naval shipbuilding industry requires heavy investment in infrastructure that might be hard to recoup with relatively sparse demand.

Buying on the open market also has advantages and disadvantages. An obvious advantage is a wider selection of designers and manufacturers, with the potential to procure world-class naval vessels. Buying in an open market can also lead to lower costs as manufacturers compete for business. Furthermore, there is less of a need to maintain indigenous design and production bases, and the Australians could simply focus on life-cycle support. Finally, it would not have to deal with the waxing and waning of demand—the so-called "valley of death" of shipbuilding inactivity that occurs when there are no major projects from the end of one period of construction to the start of another.

Our analyses suggest that *building ships in Australia carries a 30- to 40-percent price premium compared with buying the ships from foreign shipbuilders. However, that premium could drop to approximately half that level over time with a steady production program that leads to a productive workforce.* The economic benefits of a domestic shipbuilding industry are unclear and largely dependent on broader economic conditions in Australia. However, a domestic shipbuilding industry will add more than 2,000 jobs to the local economies.

In addition, building ships in Australia will minimize dependence on foreign sources and should enable and support the performance of in-service ship alterations, modernizations, and life-of-class maintenance. Supporting an Australian shipbuilding industry that is cost-effective will require specific steps, including lessening the gap between the end of the AWD program and the start of Future Frigate construction and adopting a continuous build strategy that starts a new surface combatant every 18 months to two years. There will be some challenges with replacing the *Anzac*-class ships in a timely manner, but those challenges can be overcome with careful management of the current and future fleets.

APPENDIX A

Shipbuilding in Australia: A Brief History and Current Shipyard Production Facilities

In many ways, naval shipbuilding in Australia has been a story of countervailing forces: build at home or buy abroad. Its history traces an arc that ends at a point somewhere between the two approaches: buying, from the global market, that which makes economic and operational sense (e.g., large steel ship hull structures) and supplying locally what Australia's domestic workforce and infrastructure can deliver (fabrication, installation, and integration of technical packages). All these considerations are limited by the size of the budget of a nation with a population of 22.5 million people. The points along this arc are detailed in the sections that follow in this appendix.

The Early Days

Having lost its colonies in the New World during the American Revolution, Britain sought new territory to colonize.[1] It settled on Australia and opted to use convict labor to settle the colony. In the settlement's early days, the British Royal Navy was the primary influence in governing the colony. A docking and repair station for the British navy was established at Williamstown in the 1850s, and a dry dock was subsequently opened at Mort's Dock. In 1856, the New South Wales government reserved Garden Island in Sydney Harbour as a Royal Navy ship repair site.

---

[1] The discussion in this section is largely drawn from David Stevens, *The R.A.N.—A Brief History*, website, Royal Australian Navy, undated; and Parliament of Australia, "Chapter 3: A Brief History of Australia's Naval Shipbuilding Industry," in *Blue Water Ships: Consolidating Past Achievements*, December 7, 2006a.

A dock was constructed at Cockatoo Island elsewhere in Sydney Harbour. Work started in 1847, and it took ten years to construct Fitzroy Dock. The island's Sutherland Dock was built between 1882 and 1890, and for a short time was the largest single graving dock in the world. Shipbuilding at Cockatoo began in 1870, and by World War I, more than 150 dredges, barges, and tugs had been built. Slipways were later built south of the Fitzroy Dock, and the island's biggest slipway was constructed in the Northern Shipyard in 1912. In July 1911, King George V gave the title of *Royal Australian Navy* to the Commonwealth's naval forces. Cockatoo Island became the Naval Dockyard of the Royal Australian Navy in 1913.[2]

The Cockatoo Island dockyard assembled the first Australian-built warship for RAN—the HMAS *Warrego* in June 1912. The same year, the Commonwealth government purchased the dockyard from the New South Wales government. It remained in Commonwealth ownership until 1933, when it was leased to the Cockatoo Docks and Engineering Co. Pty. Ltd.

In October 1913, formal imperial control of Australia's naval units passed to the Commonwealth Naval Board, and in that new capacity, the Australian fleet made its maiden entry into Sydney Harbour. During the same period, the Royal Australian Naval College for the training of officers was opened at Geelong, Victoria. The college moved to Jervis Bay in 1915.

World War I

When World War I broke out, the Australian Fleet consisted of a battle cruiser, six light cruisers, six destroyers, two submarines, and various support and ancillary craft. RAN operated as an element of the Royal Navy. In that role, it supported the disastrous Gallipoli campaign. Two Australian submarines were lost during the war—one sunk and one scuttled by its crew after enemy action.

---

[2] Cockatoo Island, "Sydney's Maritime History: Ship Building," Australian government, Sydney Harbour Federation Trust, undated.

Inter-War Period

With Armistice in 1918, a worldwide period of naval retrenchment began. The Commonwealth purchased the Williamstown dockyard from the Victorian government and subsequently announced a six-ship construction program there. Thereafter, the Williamstown dockyard averaged a vessel per year in addition to a substantial refit program. However, as Table A.1 shows, Williamstown was marginally active in naval shipbuilding before 1941. It was not until World War II that it again became active in building ships for RAN.

Subsequent disarmament conferences, culminating in the Washington Treaty of 1922, drastically changed naval planning. Under the terms of the treaty, the battle cruiser *Australia* was scuttled off Sydney Heads in 1924. However, additions to the battle order of the early postwar RAN included six submarines, six destroyers, and a number of sloops. All these vessels were acquired from the Royal Navy.

In the inter-war years, Australia's naval shipbuilding companies were not large enough to compete with the yards in the United Kingdom and relied on substantive foreign orders. Despite the RAN-ordered production of 22 steel ships from some Australian shipbuilding companies in the 1920s, most had to close or confine themselves to repairs. The 1930s were particularly lean for the Williamstown dockyard, which produced only three vessels.

World War II

The shipbuilding facilities at Sydney's Garden Island produced a respectable number of naval ships in the World War I and World War II eras.[3] Table A.1 lists the naval ships produced between 1912 and 1947. Several types of ships were built, mostly small to medium in size, including sloops, minesweepers, and light *River-* and *Tribal*-class escorts. Repair yard facilities also carried out extensive maintenance and refitting of ships. Anticipating war with Japan and worried about

---

[3] The name Garden Island apparently came from its use as a garden site for the ship's company of Her Majesty's Ship (HMS) *Sirius* in 1788. See Garden Island Environmental Hotline, "Captain Cook Graving Dock," web page, undated.

the security of Singapore, the British and Australian governments decided in 1939 to build a dry dock at Garden Island capable of taking the largest ships afloat. Thus, a large graving or dry dock, named for Captain James Cook, RN, was excavated and constructed there in the mid-1940s. This facility enabled Australia to accommodate very large military and commercial ships, and it obviated the need to travel 4,000 miles to Singapore, previously the nearest dry dock facility that could handle large ships. Although intended primarily for naval use (the first vessel docked was HMS *Illustrious* in February 1945), it has also been used for repairs to many civilian vessels, and so it is a crucial facility for commercial maritime enterprise in Australia.[4]

Just before the outbreak of World War II, RAN comprised two heavy cruisers, the *Australia* and the *Canberra*; two light cruisers, the *Hobart* and the *Sydney*; one destroyer, the *Voyager*; and two sloops, the *Swan* and the *Yarra*.[5] Other ships were either in reserve or deployed elsewhere.[6] Virtually all of the ships in commission were sunk, except the *Hobart* and the *Swan*, although the former did sustain serious damage from a torpedo. By 1945, the main combat strength of RAN had grown to more than 45 ships, which were supported by some 200 additional vessels, such as oilers, repair ships, auxiliary patrol vessels, and tugs.[7]

Warship construction during the war was an active industry, producing more than 30 vessels, including three *Tribal*-class destroyers and six frigates (see Table A.1). Additionally, several thousand small craft were built for RAN, the Royal Australian Air Force, the Army, and Allied forces (see Table A.2). In addition, the Australian shipbuilding industry repaired, refitted, and maintained ships not only for RAN but for foreign navies. Table A.2 lists the ships and tonnage produced.

---

[4] The Cairncross Dry Dock in Brisbane was also built during the war years. See GlobalSecurity.org, "Australian Shipbuilding Industry," March 27, 2012.

[5] Stevens (undated) lists slightly different numbers of RAN ships at the outset of the war—for example, four light cruisers rather than two, and three sloops as opposed to two. However, in each case, the overall number of ships is not large.

[6] J. H. Straczek, "RAN in the Second World War," website, Royal Australian Navy, undated.

[7] Straczek, undated.

Table A.1
Australian Naval Ship Production, 1912–1947

Year	Cockatoo Docks & Engineering Co. Limited, Cockatoo Island	Morts Dock & Engineering Co., Balmain, Sydney	Williamstown, Melbourne	Garden Island, Potts Point, Sydney
1912	HMAS *Warrego* (torpedo boat destroyer)			
1916	HMAS *Brisbane* (town light class cruiser) HMAS *Huon* HMAS *Swan* HMAS *Torrens* (*River*-class torpedo boat destroyer)			Extensive refit and repair of Allied and Australian ships during World War I
1920				Refit of British J-class submarines (J1–J5)
1922	HMAS *Adelaide* (light cruiser)			Refit of British J-class submarine (J7)
1929	HMAS *Albatross* (seaplane carrier)			
1936	HMAS *Yarra* (sloop)			
1937	HMAS *Swan* (sloop)			
1940	HMAS *Bathurst* (minesweeper) HMAS *Parramatta* HMAS *Warrego* (II) (sloop)			Work commenced on the Captain Cook Graving Dock
1941	HMAS *Bendigo* HMAS *Goulburn* HMAS *Wollongong* (minesweeper)	HMAS *Burnie* HMAS *Deloraine* HMAS *Lismore* HMAS *Lithgow* HMAS *Mildura* HMAS *Warrnambool* (minesweeper)	HMAS *Ballarat* (minesweeper)	

Table A.1—Continued

Year	Cockatoo Docks & Engineering Co. Limited, Cockatoo Island	Morts Dock & Engineering Co., Balmain, Sydney	Williamstown, Melbourne	Garden Island, Potts Point, Sydney
1942	HMAS *Arunta* HMAS *Warramunga* (*Tribal*-class destroyer) HMAS *Cessnock* HMAS *Glenelg* (minesweeper)	HMAS *Armidale* HMAS *Colac* HMAS *Dubbo* HMAS *Inverell* HMAS *Latrobe* HMAS *Wagga* (minesweeper)	HMAS *Castlemaine* HMAS *Echuca* HMAS *Geelong* HMAS *Horsham* (minesweeper) HMAS *Warreen* (survey vessel)	
1943		HMAS *Gascoyne* (*River*-class frigate)	HMAS *Benalla* HMAS *Shepparton* HMAS *Stalwell* (minesweeper)	
1944	HMAS *Barcoo* (*River*-class frigate)	HMAS *Hawkesbury* (*River*-class frigate)		
1945	HMAS *Barwan* (*River*-class frigate) HMAS *Bataan* (*Tribal*-class destroyer)	HMAS *Lachlan* (*River*-class frigate) HMAS *Macquarie* (*River*-class frigate)		Captain Cook Graving Dock officially opened

SOURCE: Parliament of Australia, 2006a, Appendix 7.

Table A.2
Australian Ship Repair During World War II

Navy	Ships	Metric Tons
Royal Australian Navy	4,008	2,150,000
Royal Navy	391	1,671,000
U.S. Navy	513	800,000
Dutch Navy	171	220,000
French Navy	44	92,000

SOURCE: Straczek, undated.

The 1960s, 1970s, 1980s, and 1990s

While World War II had fostered increased shipbuilding in Australia, production fell off at the conclusion of the war. Repair and refit work also declined, and Australia continued to buy ships from the United Kingdom. It also bought ships from the United States, including three *Perth*-class DDGs (modified *Charles F. Adams* class).

While the purchase of ships from different suppliers reflected a more discriminating approach to ship procurement, it also highlighted issues with the Australian shipbuilding industry. The *Daring*- and *River*-class destroyers built at the government-owned Williamstown and Cockatoo dockyards in the 1950s and 1960s exceeded both cost and schedule estimates. The *Daring*-class ships arrived years late and cost twice as much as the same class of ships built in the United Kingdom. The cost of the *River* class climbed by a factor of three during the project.[8]

The problems with local shipbuilding fostered a reliance on buying ships elsewhere, with the inevitable result of leaving Australian shipyards primarily to undertaking repair and maintenance work. The Williamstown yard did build two oceanographic vessels, but after the commissioning of HMAS *Torrens* in 1971, the Cockatoo Island dockyard did not build another vessel until constructing the underway replenishment ship HMAS *Success* in 1986.[9]

---

[8] GlobalSecurity.org, 2012.

[9] GlobalSecurity.org, 2012.

Frigate Project

The Australian FFG project was initiated in 1978 as a way of developing indigenous shipbuilding skills. Following the purchase of four U.S.-built *Oliver Hazard Perry*-class FFG-7 vessels, the Australian government committed to building two FFG-7 frigates at Williamstown, conditional on the dockyard committing to its ability to build the ships to RAN's requirements. In 1981, it selected HMAS *Darwin* (FFG 04) as the baseline for the build.[10] The two frigates were to be delivered between 1990 and 1994. In 1987, the Williamstown shipyard was sold to AMEC for AUD 100 million, and a contract was signed with the company extending the delivery date for the second frigate, the FFG 05, by three months. Both frigates were delivered early. Some 90 percent of AMEC's costs and 75 percent of the overall project costs were expended locally.[11] Subsequently, the dockyard was sold to Tenix, which won the contract to build eight frigates of the *Anzac* class; six were delivered to RAN, and two were delivered to the Royal New Zealand Navy.

Submarine Construction

A major event in Australian shipbuilding was the advent of the diesel-electric *Collins*-class submarine-building program.[12] The program was launched to provide a replacement for the aging *Oberon*-class vessels. Given the problems in other shipbuilding areas, many were skeptical of Australia's ability take on such a project. Intensive lobbying on the part of both industry and labor went on to build support for construction in Australia. Australian Submarine Corporation (later ASC) was established in 1985 and chosen in 1987 as the prime contractor for the design, manufacture, upgrade, and delivery of the *Collins*-class submarines.[13] Official proposals were initiated in 1978, and the development

---

[10] Parliament of Australia, 2006a.

[11] Parliament of Australia, 2006a.

[12] Nuclear power was ruled out early on because of the lack of a nuclear power industry in Australia and the political disinclination to pursue the development of such an industry.

[13] Parliament of Australia, "Chapter 4: Australian Naval Shipbuilders," in *Blue Water Ships: Consolidating Past Achievements*, December 7, 2006b.

phase of the program was approved in the 1981–1982 federal budget. After the initial bids, two companies—Ingenieur Kontor Lübeck from Germany and Kockums from Sweden—were selected to complete the bidding process for the submarine. Rockwell and Signaal were chosen for the combat system. Eventually, Kockums and Rockwell won the contracts.[14] A "greenfield" construction site (a new facility built from scratch) was built at Osborne, South Australia. The first keel was laid in early 1990 and was delivered in mid-1996, some 18 months behind schedule. The original plan called for delivery at 12-month intervals, but that did not occur. The last submarine was delivered in March 2003, some 41 months behind schedule.[15] (However, on average, the submarines were delivered some 26 months behind schedule).[16] The submarines experienced issues with their welding, periscopes, noise and vibration, propulsion system, and combat systems. Some of these issues stemmed from the optimistic expectations for the system that the existing state of technology could not deliver.[17] The *Collins*-class submarines were projected to have a 30-year hull life.

Problems notwithstanding, the construction and delivery of the six *Collins*-class submarines established that Australia could build submarines locally. Furthermore, it established ASC as a capable prime contractor for a complex shipbuilding project. The success does, however, pose somewhat of a dilemma for Australia, assuming it remains committed to diesel-electric submarines, which are increasingly unlikely to be built elsewhere in Australia. The company will have to sustain its design analysis and construction skills without a robust construction schedule. Life-cycle support for the *Collins*-class submarines will help, but sustainment skills differ significantly from a new construction program.

---

[14] Peter Jones, "A Period of Change and Uncertainty," in David Stevens (ed.), *The Royal Australian Navy*, the Australian Centenary History of Defence III, South Melbourne, Victoria: Oxford University Press, 2001.

[15] Royal Australian Navy, "*HMAS Rankin*," web page, undated.

[16] Yule, Peter, and Derek Woolner, *The Collins Class Submarine Story: Steel, Spies and Spin*, Cambridge, U.K.: Cambridge University Press, 2008, p. 325.

[17] Yule and Woolner, 2008.

The Pacific Patrol Boat Project

In 1982, the United Nations Convention on the Law of the Sea introduced a 200-nautical-mile exclusive economic zone around sovereign coastal states.[18] The sudden expansion of territorial waters from 12 to 200 nautical miles radically increased the area of ocean requiring surveillance and monitoring. This requirement created particular difficulties for the island nations throughout the southwest Pacific that now had to police an area of the ocean that was far greater than the landmass of the countries involved. Many lacked the resources, funding, and experience to take on this task. A number of Pacific nations voiced their concern about a suitable patrol force to fulfill their new requirements. The Australian government responded by instituting a Defence Cooperation Project, to provide suitable patrol vessels and associated training and infrastructure to island nations in the region.

In August 1984, the Australian government released a request for tender to construct patrol craft suitable for Pacific island nations to use in surveillance and maritime patrol operations. The Tenix Corporation won the contract to design and build the patrol boats. The first vessel, Her or His Majesty's Papua New Guinea Ship *Tarangau*, was officially handed over to the Papua New Guinea Defense Force in May 1987. In all, 22 boats were delivered to 12 countries. The Australian government sponsors follow-on support for the boats. Australia has also provided training to crewmembers. Although Australia does not operate the boats, it loans naval personnel to each nation operating the boats to provide technical advice.

Australian Defence Industries

In May 1989, Australian Defence Industries (ADI) was created as a government-owned corporation to take over the major defense industry facilities that remained under government ownership. It has four operating divisions: naval engineering at the Garden Island dock-

---

[18] This discussion is largely drawn from Royal Australian Navy, "The Pacific Patrol Boat Project, *Semaphore: Newsletter of the Sea Power Centre—Australia*, No. 2, Canberra, Australia: Department of Defence, February 2005.

yard, ammunition and missiles, weapons, and engineering and military clothing. The idea was that ADI was part of a broader process to make government factories and shipyards an integral part of Australian industry. ADI's key project was as the prime contractor for the *Huon*-class minehunter. The corporation won the AUD 917 million project in 1994 to build six minehunter coastal vessels, five of which it built completely at Carrington in Newcastle. All six were delivered on time.[19] ADI was privatized in 1999 and later, in 2006, was taken over by Thales Australia. Thales Australia operates as an independent commercial entity managing and operating the dry dock at Garden Island in Sydney. It shares the Garden Island facility with RAN.

2000s

By the mid-2000s, the structure of the Australian shipbuilding industry had changed, with Thales running the Garden Island dockyard in Sydney and BAE Systems running the Williamstown dockyard. The industry was quite capable of building frigates, such as the *Anzac* class, which displaced about 3,600 metric tons and were about 110 m long.[20] The *Anzac* class replaced the *River*-class destroyer escorts (~2,200 metric tons). Larger ships had been constructed, such as the *Durance*-class oiler (HMAS *Success*) at about 18,000 metric tons, but hulls larger than 25,000 metric tons were beyond the capacity of the local shipbuilding industrial facilities.[21]

---

[19] Parliament of Australia, 2006b, paras. 4.37–4.39.

[20] IHS (undated) and other publicly available sources.

[21] However, that does not exclude Australia from playing a meaningful and profitable role in ship construction. It does mean that large steel hulls are not its preferred focus, nor necessarily an arena in which the Australian shipbuilding industry wishes to compete. A more recent approach has been to purchase the basic hull offshore and install subsystem infrastructure in Australia. This approach appears to minimize Australia's weakness in constructing large steel hulls and play to its strength of developing technology packages, which also has some spillover potential to other industries.

Armidale-Class Patrol Boat Replacement Project

The *Armidale* class of patrol boat replaced the earlier fleet of 15 *Fremantle*-class boats.[22] The project was initially begun as a joint OPV with the Royal Malaysian Navy, but it was not successful. The restarted new (*Armidale*) patrol boat project began in 1999. The first vessel, the HMAS *Armidale*, joined the fleet in 2005. A total of 14 vessels were acquired.

The ships are operated by RAN as the Australian Patrol Boat Force Element Group, based in Cairns and Darwin. They carry out border protection, fishery patrol, and interceptions of unauthorized vessels. Two vessels protect oil and gas facilities of the North West Shelf Venture.

The *Armidale*-class boats are larger and heavier than their predecessors and have greater range. They are also assigned multiple crews so that they can spend more time at sea without interfering with crew rest or training. The design has an aluminum hull and a top speed of 25 knots.[23]

All 14 boats were built by Austral at its Henderson shipyard in Western Australia. Deliveries in addition to those described above included six in 2006, five in 2007, and the final vessel in February 2008.

Production Facilities and Phases Associated with Ship Production Today

Production of a naval surface ship today involves numerous facilities, including a wide range of shops, cranes, docks, piers, and specialized equipment. Producers employ facilities at different times, in different sequences, and in different ways, depending on the platform being built, yard organization or layout, build strategy, and many other factors. Figure A.1 depicts the three production phases that we refer to in this analysis, with each phase corresponding to the particular facilities it requires. These are not the traditional shipbuilding phases commonly

---

[22] This discussion is drawn from IHS (undated) and other publicly available sources.

[23] The ships have a range of 3,000 miles at 12 knots. Their main weapon is a 25-mm Bushmaster cannon, and they also mount two .50-caliber machine guns.

known and used throughout the industry (i.e., design, production, outfitting, test, commissioning, and trials). Nevertheless, we defined these phases because the traditional categorizations did not allow us to adequately map the use of particular facilities. Figure A.1's purple area corresponds to the *pre-final assembly* phase, the yellow to the *final assembly* phase, and the red to the *afloat outfitting* phase.

The pre-final assembly phase entails a manufacturing period before final assembly of blocks and modules begins and before the ship occupies an assembly location. During this period, facilities such as pipe and steel fabrication shops, unit assembly areas, and lay-down areas are used. Final assembly begins when a producer starts assembling the ship, using a facility such as a dry dock, floating dock, slipway, land-level area, or ship assembly hall. Afloat outfitting begins when a ship is launched or floated and ends when the ship is delivered. A ship in the afloat outfitting phase would require a pier, quay, lock, or dock.

There is some overlap in the use of different facilities throughout each phase. In many cases, certain facilities—cranes, shops, or fabrication facilities that are associated with the pre-final assembly—are used throughout the final assembly and afloat outfitting phases. Generally, the final assembly and afloat outfitting phases are mutually exclusive, but sometimes a final assembly facility will be used for outfitting.

Figure A.1
Ship Production Phases

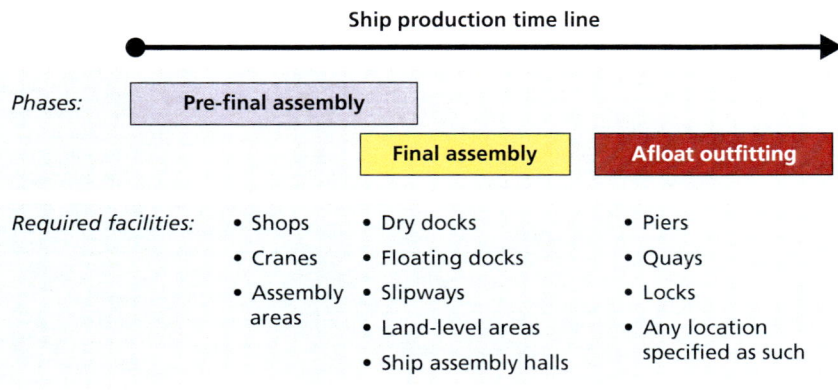

In this analysis, we focused our evaluation of facility throughput on the facilities needed for final assembly and afloat outfitting (the last two phases we defined) for two main reasons:

1. There were not many consistent measures of throughput for the other types of facilities. For facilities such as shops, it is very difficult to devise objective measures of throughput. The number of pipes that a pipe shop can manufacture, for example, depends on the complexity of each unit, length, diameter, number of bends, and so on. Thus, simply stating the number of pipes per day as capacity could be misleading.
2. Consistent measures of throughput would require a prohibitive amount of data from the shipyards. For the example of a crane, the throughput depends on where the crane is located in the yard, what the build strategy is of the program that will utilize that crane, how easily the crane can be moved, and how easily supplemental cranes can be brought in. Each unit lift must be tracked and assessed. This information would be needed for each crane, for each lift, and for each vessel in the yard. Such a data collection was beyond the scope and means of this study and would have placed an undue burden on the shipyards providing information.

APPENDIX B

Shipbuilding Model and Assumptions

A shipbuilding industrial base can be thought of as a system that produces a set of outputs for a given set of inputs. The outputs of the shipbuilding system are ships, and broadly speaking, there are three main inputs: ship requirements, a workforce, and labor rates. Ship requirements include the number and type of vessels required; these force structure requirements translate to demands for different quantities and types of skilled labor. Important workforce variables include the size of the workforce, its composition in terms of people with relevant shipbuilding skills, the capacity of the workforce to grow or contract over time, and the productivity of that workforce, among other variables. The effectiveness of the shipbuilding system can be measured in terms of the cost and time it takes to produce ships demanded; the cost depends on labor rates.

As discussed in Chapter Four, in prior research, RAND developed a Shipbuilding and Force Structure Analysis Tool that represents the relationships between these key variables and serves as a model for a shipbuilding industrial base.[1] The present study adapted that prior model to assess the Australian shipbuilding industrial base under different futures. This appendix documents the assumptions that were made to adapt and employ the RAND tool to model Australian circumstances.

---

[1] Arena, Schank, and Abbott, 2004.

Shipbuilding Workforce Framework

Shipbuilding is a complex endeavor that requires people with a range of different skills. At the highest level of abstraction, the range of demanded skills includes managers; technical workers, such as designers and engineers; and people experienced with manufacturing. Within these broad categories, there are many specialties, such as administrators, cost estimators, welders, pipe fitters, and crane operators, to name just a few. Different shipyards sometimes define skills differently, so we developed a standardized framework for shipbuilding skills to facilitate modeling the demand for and supply of shipbuilding labor in Australia.

Table B.1 shows the framework developed for this study, which consists of three levels—category, subcategory, and specific skill. The highest level distinguishes management and technical workers from manufacturing. At the second level, within general management and technical, the framework distinguishes the category of *general management*, which includes executives, administrators, and marketers, from the category of *technical*, which includes designers, engineers, cost estimators, and the like. Within manufacturing, the framework distinguishes workers manufacturing structural elements (*structure*), those contributing to ship outfitting (*outfitting*), and those that are providing general support, such as riggers and crane operators (*direct support*). The shipbuilding industrial model separately represents the demand and supply of shipbuilding labor at the second level of this hierarchy, focusing on five broad skill categories: general management, technical, structure, outfitting, and direct support. Our assumption is that this level of abstraction includes the full breadth of skills required in shipbuilding, is sufficiently detailed to understand the range of different demands on a shipbuilding workforce, and is sufficiently simple to make the analysis tractable. The third level of the hierarchy provides the specific skills required for shipbuilding.

Table B.1
Shipbuilding Workforce Framework

Category	Subcategory	Specific Skill
General management and technical	General management	Management
		Administration
		Marketing
		Purchasing
	Technical	Design
		Drafting/CAD specialist
		Engineering
		Estimating
		Planning
		Program control/project management
Manufacturing	Structure	Steelworker, plater, boilermaker
		Structure welder
		Shipwright/fitter
		Team leader, foreman, supervisor, progress control (fabrication)
	Outfitting	Electrician, electrical tech, calibrator, instrument tech
		HVAC installer
		Hull insulator
		Joiner, carpenter
		Fiberglass laminator
		Machinist, mechanical fitter/tech, fitter, turner
		Painter, caulker
		Pipe welder
		Piping/machinery insulator
		Sheet metal
		Team leader, foreman, supervisor, progress control (outfitting)
		Weapon systems
	Direct support	Rigger, stager, slingers, crane, and lorry operators
		Service, support, cleaners, trade assistant, ancillary
		Stores, material control
		Quality assurance/control

Summary of Key Variables

Table B.2 defines the key variables in this study. The remainder of this appendix describes the assumptions that our analysis made about each variable or the different configurations of the variables that are explored.

Shipbuilding Demand Variables

Acquisition Scenario

As introduced in Table 3.2, the acquisition scenario represents the number and type of ships to be manufactured and their intended start and delivery dates. The acquisition scenario is perhaps the most important variable in the analysis, and the analysis explores the effect of this variable on the industrial base. All explorations are variants of two acquisition plans articulated by the White Paper team, which are articulated in Table B.3.

In addition to these baseline acquisition scenarios that reflect notional requirements of RAN, we also explored the potential addition of three to five OPVs for the explicit purpose of lessening the production gap between the completion of the AWD program and the start of the Future Frigate program. These were not associated with any specific acquisition scenario.

Production Plans

The production plan is the number and type of ships or ship blocks assigned to different shipyards to realize an acquisition plan. We consider one baseline production plan and one or more variants for each ship class.

Future Frigate

There is one baseline production plan for the construction schedule of the Future Frigate and multiple excursions that were examined in our sensitivity analysis. In this baseline, production of the first-of-class begins in 2020, with deliveries commencing in 2026. The following seven ships' construction extends into the mid-2030s. Construction of

Table B.2
Summary of Key Variables

Variables	Factor	Description
Shipbuilding demand	Acquisition scenario	Number and type of ships to be manufactured and their intended start and delivery dates
	Production plan	Number and type of ships or ship blocks assigned to different shipyards to realize an acquisition plan
	Demand profiles	Quantity and type of workers required to construct a single ship over time
	Unit learning curve	Relative man-hours required to construct an additional ship compared with the number of man-hours required to produce the previous ship
Shipbuilding workforce	Initial labor	Quantity, type, and experience of available labor in the first simulated year
	Labor pool	Quantity and type of fully experienced workers available for hiring without training costs
	Hiring rate	Number of new workers that could be added to the workforce in a given quarter as a percentage of the size of the workforce in that quarter
	Firing rate	Number of workers that can be made redundant in a given quarter as a percentage of the size of the workforce in that quarter
	Workforce floor	Minimum number of workers that a shipyard must retain in a given skill category, regardless of demand
	Workforce ceiling	Maximum number of workers that a shipyard can sustain in a given skill category, regardless of demand
	Productivity	Proficiency of worker, expressed relative to a 100-percent fully proficient worker (in this study, we will model productivity as a function of experience)
	New hire distribution	Distribution of new hires as a function of experience (in this study, we will model seven years of experience)
	Mentoring ratio	Number of new workers that can be hired for every one fully experienced worker
Shipbuilding cost	Direct labor rate	Average direct hourly rate for one FTE worker in a given skill category
	Overhead rate	Percentage increase in hourly rate to account for fixed and variable overhead costs (in this study, we allow overhead rate to vary as a function of size of business base)
	Training cost	Quarterly training cost for workers, as a function of skill category and experience
	Termination cost	Cost of making a single worker redundant, as a function of years on the job
	Hiring cost	Cost of hiring a worker
	Maximum overtime	Maximum amount of overtime that may be worked in a given quarter

Table B.3
Summary of Baseline Acquisition Scenarios

Scenario	Ship Class	Quantity	Delivery Date(s)
Scenario 1	Future Frigates	8, replacing the *Anzac* class one for one[a]	2026–2035
	Hobart-class destroyers	3, with possible addition of 1 to lessen production gap	2016–2019
	Patrol boats	14, replacing *Armidale*-class patrol boats	2021–2026
	Littoral multirole vessels	21	2035–
Scenario 2	Future Frigates	8, replacing the *Anzac* class one for one	2026–2035
	Hobart-class destroyers	3, with possible addition of 1 to lessen production gap	2016–2019
	Patrol boats	8	2021–2026
	Littoral multirole vessels	21	2026–

[a] The original White Paper scenario posited eight to ten Future Frigates. For purposes of analysis, we assume eight Future Frigates, in alignment with the number of *Anzac*-class frigates that they will replace.

Hulls 1 and 2 is spaced by three years, while construction of Hulls 2 and 3 is spaced by two years. Following Hull 3, a new ship begins each year through the end of the class.

The alternative production plans considered in our excursions change two parameters on the baseline. First, we consider variations on when the Future Frigate begins construction, in either the first quarter of 2020 (the baseline), the first quarter of 2018, the fourth quarter of 2018, or the first quarter of 2017. Second, we vary the drumbeat, or rate at which Hulls 3 through 8 are produced, considering cases where frigates are produced once every year (base case), once every 1.5 years, and once every two years. In all of these variants, we assume that the delivery dates shift accordingly.

The same basic production plans are considered for both cases, where the industrial base is responsible for all of construction (full capability path) or only final outfitting (limited capability path). However, for the outfitting-only case, we assume that ship outfitting starts two years after the start of ship construction, reflecting the fact that approximately the first two years would be spent constructing the hull overseas.

This baseline plan is shown in Table B.4. The alternative plans are straightforward modifications of this plan per the details described here.

Patrol Boat

In the baseline production plan for the patrol boats, we assume that construction begins in 2020 and results in 14 patrol boats before the end of construction in 2026. Two boats would be built per year. We consider one excursion where construction of the patrol boats begins in 2017. Further variants of each of these production plans can be modified to accord with acquisition Scenario 2, where the number of patrol boats is reduced to eight.

The baseline plan is shown in Table B.5. The alternative plans are straightforward modifications of this plan per the details described here.

Littoral Multirole Vessel

The LMRVs represent ships that would fit somewhere in size between the Future Frigates and the patrol boats. In acquisition Scenario 1, we assume that the LMRV ship program, if developed, begins construction in 2033 and extends through the mid-2040s, resulting in 21 ships. We consider one excursion where construction of the LMRV begins in 2026. Table B.6 shows the baseline production plan.

Table B.4
Future Frigate Construction Schedule (Base Case: 2020 Start, One-Year Drumbeat)

Future Frigate Hull #	Start Quarter	Start Year	End Quarter	End Year
FF 1 (first of class)	1st	2020	2nd	2026
FF 2	1st	2023	2nd	2028
FF 3	1st	2025	2nd	2030
FF 4	1st	2026	2nd	2031
FF 5	1st	2027	2nd	2032
FF 6	1st	2028	2nd	2033
FF 7	1st	2029	2nd	2034
FF 8	1st	2030	2nd	2035

Table B.5
Patrol Boat Construction Schedule (Base Case: 2020 Start)

Patrol Boat Hull #	Start Quarter	Start Year	End Quarter	End Year
PB 1	1st	2020	2nd	2021
PB 2	3rd	2020	4th	2021
PB 3	1st	2021	2nd	2022
PB 4	3rd	2021	4th	2022
PB 5	1st	2022	2nd	2023
PB 6	2nd	2022	3rd	2023
PB 7	3rd	2022	4th	2023
PB 8	1st	2023	2nd	2024
PB 9	2nd	2023	3rd	2024
PB 10	3rd	2023	4th	2024
PB 11	1st	2024	2nd	2025
PB 12	3rd	2024	4th	2025
PB 13	1st	2025	2nd	2026
PB 14	3rd	2025	4th	2026

Air Warfare Destroyer

The three planned *Hobart*-class AWDs are included in the model as a single profile. The program began in 2007 and is planned to end in 2019. Table B.7 shows the construction schedule for the AWD. The analysis explores one variant of adding a fourth AWD, with production beginning in 2017; see Table B.8 for the plan that includes a fourth AWD.

Offshore Patrol Vessel

The three to five OPVs introduced for the express purpose of lessening the gap in workforce demand were planned to begin construction at times that would best optimize lessening in the gap. Table B.9 shows the production plans for these cases.

Table B.6
Littoral Multirole Vessel Construction Schedule
(Base Case: 2033 Start)

Littoral Multirole Vessel Hull #	Start Quarter	Start Year	End Quarter	End Year
LMRV 1	2nd	2033	1st	2035
LMRV 2	4th	2033	3rd	2035
LMRV 3	2nd	2034	1st	2036
LMRV 4	4th	2034	3rd	2036
LMRV 5	2nd	2035	1st	2037
LMRV 6	4th	2035	3rd	2037
LMRV 7	2nd	2036	1st	2038
LMRV 8	4th	2036	3rd	2038
LMRV 9	2nd	2037	1st	2039
LMRV 10	4th	2037	3rd	2039
LMRV 11	2nd	2038	1st	2040
LMRV 12	4th	2038	3rd	2040
LMRV 13	2nd	2039	1st	2041
LMRV 14	4th	2039	3rd	2041
LMRV 15	2nd	2040	1st	2042
LMRV 16	4th	2040	3rd	2042
LMRV 17	2nd	2041	1st	2043
LMRV 18	4th	2041	3rd	2043
LMRV 19	2nd	2042	1st	2044
LMRV 20	4th	2042	3rd	2044
LMRV 21	2nd	2043	1st	2045

Table B.7
Air Warfare Destroyer Construction Schedule
(Existing Program)

Air Warfare Destroyer Hull #	Start Quarter	Start Year	End Quarter	End Year
AWD 1–3	1st	2007	4th	2019

Table B.8
Air Warfare Destroyer Construction Schedule (Adding a Fourth Hull)

Air Warfare Destroyer Hull #	Start Quarter	Start Year	End Quarter	End Year
AWD 4	1st	2017	2nd	2022

Table B.9
Offshore Patrol Vessel Construction Schedule

Construction Case	Offshore Patrol Vessel Hull #	Start Quarter	Start Year	End Quarter	End Year
3 OPVs (full build)	OPV 1	1st	2018	4th	2020
	OPV 2	4th	2018	3rd	2021
	OPV 3	3rd	2019	2nd	2022
4 OPVs (full build)	OPV 1	1st	2017	4th	2019
	OPV 2	4th	2017	3rd	2020
	OPV 3	1st	2019	4th	2021
	OPV 4	2nd	2020	1st	2023
5 OPVs (full build)	OPV 1	2nd	2017	1st	2020
	OPV 2	4th	2017	3rd	2020
	OPV 3	3rd	2018	2nd	2021
	OPV 4	3rd	2019	2nd	2022
	OPV 5	2nd	2020	1st	2023

Demand Profiles

Demand profiles define the quantity and type of labor required per quarter to develop a single ship. The acquisition plans under investigation in this study require developing demand profiles for the following ships:

- Future Frigate: 5,500 to 8,000 metric tons
- Patrol boat: 300 to 350 metric tons
- LMRV: 1,000 to 1,800 metric tons
- AWD: as designed
- OPV: 1,700–1,800 metric tons.

Each ship built will result in a unique profile, because no single ship is ever the same as another ship. However, for the purposes of modeling, we define a standard profile for each ship type or each ship class. Profiles were developed by subject-matter experts, informed by surveys collected in Australia from current shipbuilding companies. Using this information, we developed profiles for each of the programs analyzed in this report. The length of construction was defined using the same data sources. Table B.10 summarizes our assumptions that we now describe in more detail.

Future Frigate—Full Capability Path

Neither the conceptual or detailed design of the Future Frigate has been determined, so it is uncertain how large or complicated it will be and thus how much effort it will take to build it. The range of potential designs includes ships simpler than the AWD, similar in complexity to the AWD, and more complex than the AWD. Without detailed plans and designs for the Future Frigate, using precise estimates would be inappropriate. Thus, we considered one base case and several excursions for the Future Frigate demand profile to reflect a range of possibilities.

Table B.10
Summary of Demand Profile Assumptions

Ship Class	Total Man-Hours	Duration (Quarters)
Future Frigate (full capability path)	1,800,000	16
	3,000,000	20
	5,000,000 (base case)	22
	7,000,000	24
Future Frigate (limited capability path)	750,000	11
	1,500,000	14
	2,500,000 (base case)	18
	3,500,000	22
Patrol boat	140,000	5
Littoral multirole vessel	500,000	8
Air warfare destroyer (new hull)	5,524,000	22
Offshore patrol vessel	700,000	12

As a baseline, the first-of-class Future Frigate is assumed to take an estimated 5.5 million man-hours. Follow-on ships will begin at 5 million man-hours, and learning will occur through the program as more ships are built. Building the first-of-class ship is estimated to take 26 quarters, with each follow-on ship estimated to take 22 quarters. Figure B.1 shows the profile for the first-of-class Future Frigate, and Figure B.2 shows the profile for the first of the follow-on ships.

We consider three excursions, where the total level of effort for constructing the Future Frigate is 1.8 million man-hours (16 quarters), 3 million man-hours (20 quarters), or 7 million man-hours (24 quarters), respectively. In these excursions, we assume that the distribution of effort over time remains the same; we merely scale the total level of effort and duration.

For each case (baseline or excursion), the first-of-class hull is assumed to require 10 percent more man-hours and take four quarters longer than the specified effort and duration required. For example, for the 3 million man-hour case, the first-of-class hull is assumed to require 3.3 million man-hours and take 24 quarters to complete. These

Figure B.1
Workforce Profile for First-of-Class Future Frigate (Full Capability Path)

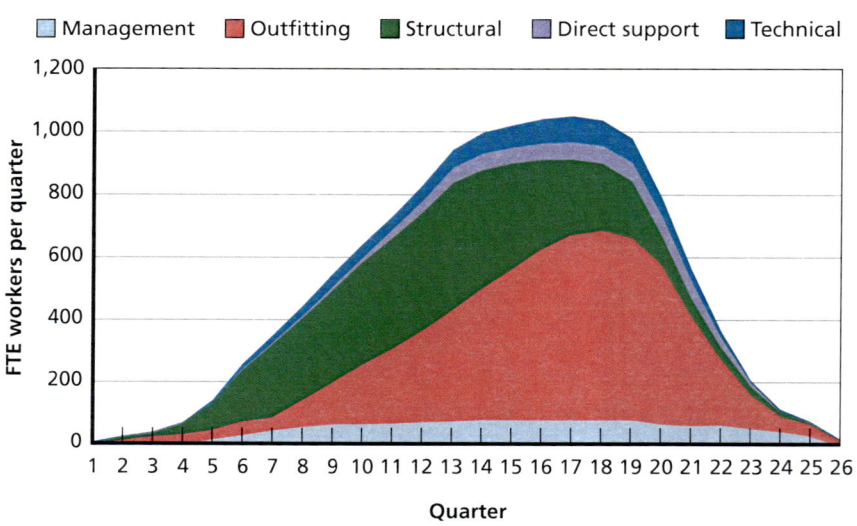

Figure B.2
Workforce Profile for First Follow-On Future Frigate (Full Capability Path)

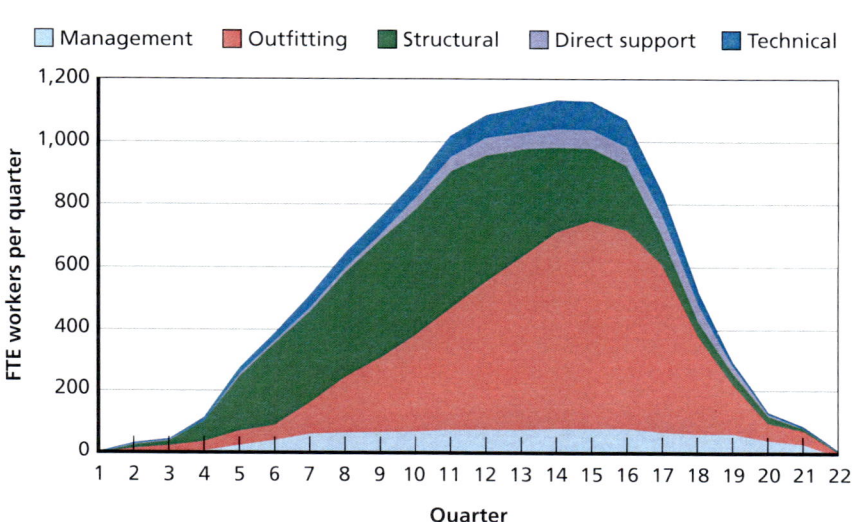

demand profiles thus look comparable to those for the baseline, except the FTE per quarter is scaled proportionally.

Future Frigate—Limited Capability Path

In the limited capability path (Path 2), the HM&E equipment of the Future Frigate will be built outside Australia, and Australia will outfit the ship. This type of build is different, and therefore a different profile and labor amount is needed. Nonetheless, there is the same uncertainty about the basic design requirements, requiring us to explore a range of possibilities.

For this, we use the shipbuilding profile for the BAE-built LHD as a starting point for the distribution of work across skill categories over time. The LHD was mostly built in Spain and outfitted in Australia. Therefore, we assume that the LHD profile is a reasonable representation of the phasing of work that would be needed.

In the base case, the amount of labor is assumed to be 2.5 million man-hours, and the duration is assumed to be 18 quarters, as outfitting is the primary goal, with less emphasis on structural work and smaller amounts of labor required for management, support, and technical skills. Figure B.3 shows the single ship workforce demand profile for the case of outfitting only.

We considered three variants, where the total level of effort for outfitting a Future Frigate otherwise constructed abroad was 750,000 man-hours (11 quarters), 1.5 million man-hours (14 quarters), and 3.5 million man-hours (22 quarters). As with the full-build case, we assume that the distribution of work across skill categories is the same, and we merely scale to account for the variation in the total level of effort and duration. Thus, the demand profiles look comparable to those for the base case, except the FTE per quarter is calculated proportionally to the alternative level of effort. Unlike the full-build scenarios, no adjustment was made to reflect inefficiencies for a first-of-class build.

Patrol Boat

An estimated 140,000 man-hours are needed for each patrol boat, while construction lasts 5 quarters. Figure B.4 depicts the assumed patrol boat workforce demand profile.

Figure B.3
Workforce Profile for One Outfitting-Only Future Frigate (Limited Capability Path)

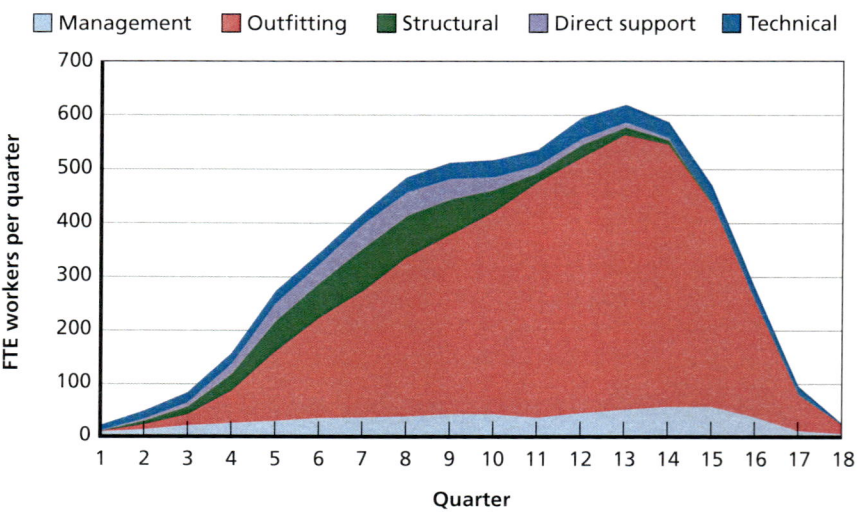

Figure B.4
Workforce Profile for One Patrol Boat

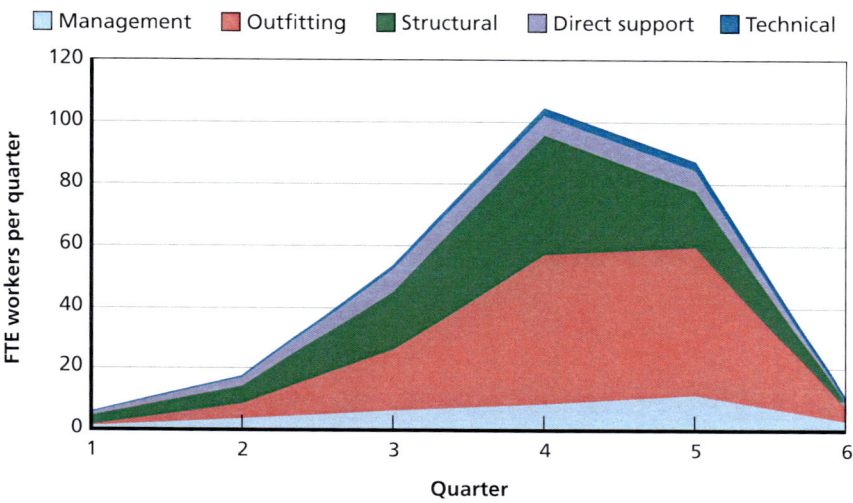

Littoral Multirole Vessel

At 1,000 to 1,800 metric tons, the LMRV is larger than a patrol craft but not quite the size of a surface combatant, such as the AWD and Future Frigate. We assume 500,000 man-hours across eight quarters are required to construct each LMRV. Figure B.5 shows the assumed LMRV workforce demand profile.

Air Warfare Destroyer—Existing Program

The AWD profile was developed from data provided by the shipbuilders, reflecting their actual and planned production starting in 2014. The production plan provided by the shipbuilders' surveys was stretched to account for the revised delivery dates announced by the Australian government in December 2014 (the specific delivery dates were provided to RAND by the Australia White Paper team). The workforce profile for AWD past and future construction (Hulls 1–3) is shown in Figure B.6.

Figure B.5
Workforce Profile for One Littoral Multirole Vessel

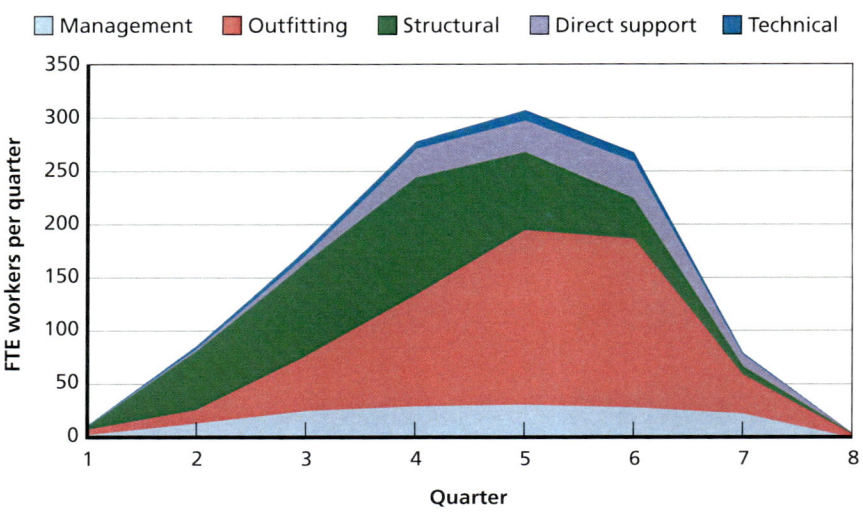

RAND RR1093-B.5

Figure B.6
Workforce Profile for Existing Air Warfare Destroyer Program

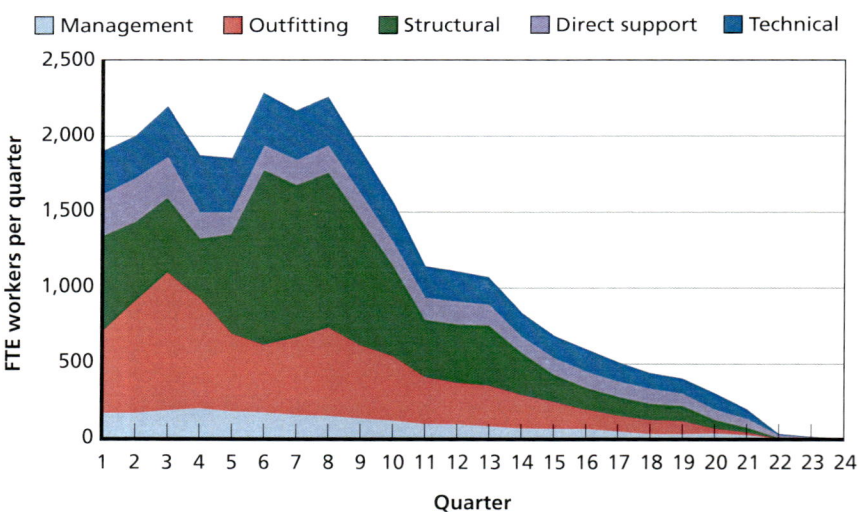

NOTE: Demands for the existing AWD program are assumed to start in the first quarter of 2014.

RAND RR1093-B.6

Air Warfare Destroyer—Adding a Fourth Hull

One scenario includes adding a fourth AWD; therefore, a single ship profile was developed. It was adapted from the existing AWD Hull 3. AWD Hull 4, should it occur, is estimated to require 5.5 million man-hours to construct over 22 quarters. This is similar to the first three AWDs. Nonrecurring engineering for the fourth AWD is assumed to be unnecessary, as it was already included in the profiles for AWD Hulls 1–3. Figure B.7 shows the assumed workforce demand profile for a fourth AWD.

Offshore Patrol Vessel

The OPV is a boat similar to the patrol boat, detailed above. The OPV is larger, and therefore the labor hours and construction duration are larger. The OPV case was developed to lessen the workforce demand gap between construction for the AWD and the Future Frigate. This OPV is assumed to take 700,000 man-hours and 12 quarters to produce. The workload profile was based on that used for the Future Frigate, though fewer man-hours are estimated for OPV production

Figure B.7
Workforce Profile for a Fourth Air Warfare Destroyer

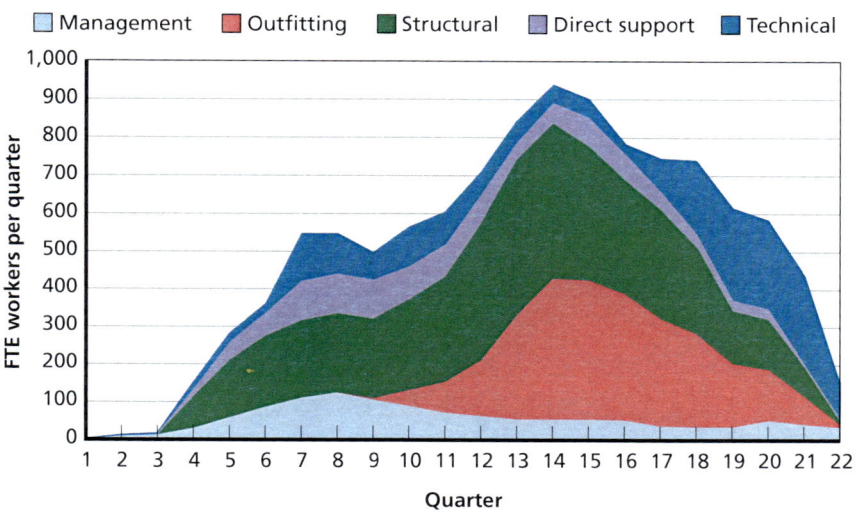

because it is a smaller and simpler ship. Figure B.8 depicts the assumed workforce demand profile. Additional scenarios involving OPVs are considered in Appendix D.

Unit Learning Curve

For all ship programs in this study, we model the reduction in production hours based on experience using a unit learning curve. The slope of the unit learning curve represents the reduction in hours when the quantity of ships of a class produced doubles. So for a 95-percent learning slope, the second ship's hours are 5 percent lower than the first ship, and by the fourth ship, the hours are 5 percent lower than the second ship. For most programs, we assume a 95-percent unit learning curve. The exception to this assumption is related to the Future Frigate program, where we examine both a 90- and 95-percent learning slope. Also, as discussed earlier in this appendix, we assumed a fixed stepdown of 10 percent between the Future Frigate's first and second hull, regardless of the learning slope.

Figure B.8
Workforce Profile for One Offshore Patrol Vessel

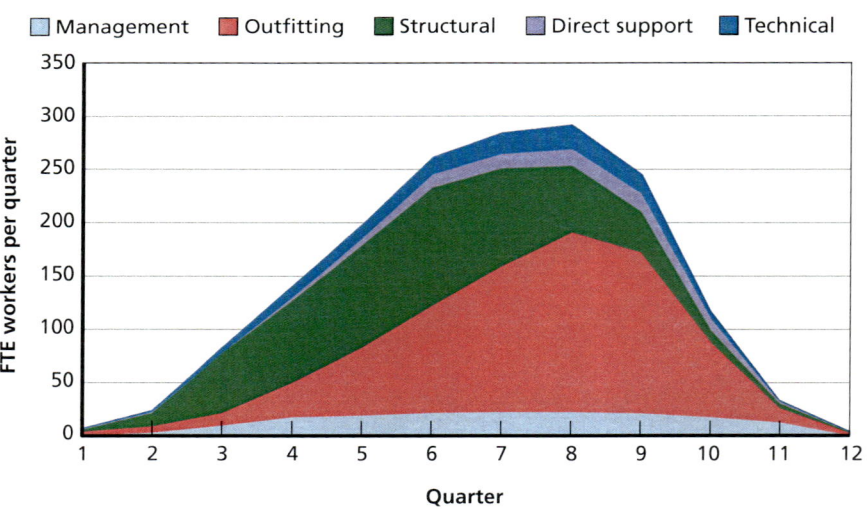

Shipbuilding Workforce Variables

Initial Workforce
Initial workforce is the workforce available in the first quarter of the simulation (the first quarter of 2014). We assume that initial labor meets initial demand exactly with fully productive workers.

Labor Pool
The starting labor pool consists of fully productive workers. The size of the pool is defined as the difference between the maximum and minimum planned demand from 2014 to 2016. This assumption was made to smooth out any large variations in demand through the end of the initial AWD program. This pool is nearly exhausted by 2017, after which work on other projects begins.

Hiring Rate
The hiring rate is the number of new workers that could be added to the workforce in a given quarter as a percentage of the size of the workforce in that quarter. We assume that the hiring rate is 10 percent on a per-skill-category basis, unless we explicitly treat it as a variable. In our sensitivity analysis, we explore a variation of 20-percent and 30-percent hiring rate. In general, this variable is driven by the labor market conditions and accordingly may vary from shipyard to shipyard.

Firing Rate
The firing rate is the number of workers that can be made redundant in a given quarter as a percentage of the size of the workforce in that quarter. We assume that a shipyard may fire up to 100 percent of its workers each quarter.

Workforce Floor
The workforce floor is the minimum number of workers that a shipyard must retain without firing, irrespective of demand. Unless the analysis is explicitly treating the workforce floor as a variable, we assume the workforce floor in a given quarter is 5 percent of peak future demand (within each skill category). This assumption is necessary to prevent

an artificial circumstance where the workforce goes to zero, leaving no basis from which to grow.

Workforce Ceiling

The workforce ceiling is the maximum number of workers that a shipyard will hire by policy; it does not reflect the capacity of the labor market. For a baseline assumption, we assume that the shipyard workforce ceiling is equal to 100 percent of peak planned demand (within each skill category). In our sensitivity analysis, we explore a variation that the workforce could reach up to 120 percent of peak planned demand (within each skill category).

The workforce ceiling comes into play when the shipyard falls behind on planned work; the assumption is that the shipyard would *not* hire unlimited new workers just to catch up. Rather, the shipyard would hire, at most, limited numbers in excess of peak planned demand to avoid the costs of hiring and terminating workers needed only temporarily.

Productivity

Productivity is the proficiency of a worker, expressed relative to a 100-percent fully proficient worker. In this study, we model productivity as a function of experience. Table B.11 shows the assumed productivity, informed by shipyard surveys and prior RAND shipbuilding research. We assume that workers gain experience on an annual basis. In our sensitivity analysis, we explore a variation in which workers gain proficiency twice as quickly. For example, in comparison with Table B.10, a management employee with no experience will have 30-percent productivity, and a management employee with one year of experience will have 80-percent productivity.

A further assumption is that productivity can only increase, not decrease, for an individual worker. So, in a transition period, workers who are maintained to sustain a workforce level will remain fully productive, regardless of whether they have work to sustain their skills. In practice, productivity levels will need to be earned. Workers will need to be given real shipbuilding work to sustain the experience levels during the transition. This will be a challenge of dealing with the gap.

Table B.11
Productivity as a Function of Experience, by Subcategory

Experience	Productivity (%)				
	Management	Outfitting	Structure	Direct Support	Technical
None	15	11	23	15	15
1 year	40	36	55	33	40
2 years	65	66	80	55	64
3 years	85	87	91	75	85
4 years	95	96	97	93	95
5+ years	100	100	100	100	100

New Hire Distribution

The new hire distribution is the distribution of new hires as a function of experience, which governs the productivity of new hires. Table B.12 shows how new hires are distributed by experience levels, informed by shipyard surveys and prior RAND shipbuilding research.

Mentoring Ratio

Mentoring ratio is the number of new workers that can be hired for every one fully experienced worker; it is one potential limit on hiring new workers. For each fully experienced worker at the yard, some number of new, not fully productive hires may be brought in. We assumed a mentor ratio of 4:1, meaning four new hires can be employed for every fully productive worker. The number of fully productive workers is the sum of fully productive workers employed at the yard and hires from the fully experienced labor pool.

Shipbuilding Cost Variables

Direct Labor Rate

The direct labor rate is the average hourly wage of shipyard labor, not accounting for overhead, by skill category. All surveyed shipyards indicated that their rates were proprietary and should not be disclosed to the government. Table B.13 shows the generic labor rates assumed in this study, as informed by shipyard surveys and the Australian Bureau of Statistics.

Table B.12
New Hire Distribution

Experience (years)	Percentage of New Hires with This Much Experience
None	50
1	20
2	20
3	10

Table B.13
Summary of Direct Labor Rates, by Subcategory

	Direct Labor Rate (AUD)				
	Management	Outfitting	Structure	Support	Technical
FTE hourly rate	61	40	40	40	53

Overhead Rate

Overhead rate represents the percentage increase in hourly rate to account for fixed and variable overhead costs. In this study, we allow the overhead rate to vary as a function of the size of the business base. In general, overhead structure varies shipyard to shipyard. All Australian shipyards indicated that overhead rates were proprietary and not to be disclosed to government. Thus, this study employs a generic overhead structure.

We consider two different overhead rate models for the separate cases of a fully capable shipbuilding industrial base (full capability path) and an industrial base focused on outfitting only (limited capability path). Our assumption is that the industrial base would restructure after the AWD program if the Australian government decided to support only outfitting. Table B.14 summarizes the overhead rate assumed for the fully capable industrial base and in the outfitting-only industrial base prior to the conclusion of the AWD program in 2019. Table B.15 summarizes the overhead rates assumed for the outfitting-only industrial base after the conclusion of the AWD program in 2019.

Table B.14
Generic Overhead Rate Assumptions (Full Capability Path)

Demand (FTE workers)	Overhead Rate (%)
< 250	200
500	200
750	162
1,000	130
1,250	111
1,500	98
1,750	89
2,000	82
2,250	80
> 2,500	80

Table B.15
Generic Overhead Rate Assumptions (Limited Capability Path)

Demand (FTE workers)	Overhead Rate (%)
<250	200
350	175
450	144
550	124
650	110
750	100
850	92
950	86
1,050	81
>1,150	80

Training Cost

Training costs are the costs to train a new hire to be fully productive. We assume that the training costs vary by experience level of the new hire. The specific costs are summarized in Table B.16.

Termination Cost

Termination costs are the costs paid when making a worker redundant, which in Australia varies by years of service. These costs vary by shipyard and are generally proprietary, although the Australian government sets certain limits. Table B.17 summarizes the assumed termination costs as a percentage of annual salary.

Hiring Cost

We assume that additional costs associated with hiring new workers are included in the overhead rates.

Maximum Overtime

The maximum overtime rate is the maximum amount of overtime that can be worked in a given quarter. We assume a maximum overtime rate of 10 percent (within each skill category). Further, we assume that all overtime is equally productive and that overtime does not cost more than on a per-hour basis.

Table B.16
Training Costs

Experience (years)	Annual Cost per FTE Worker (AUD)
None	10,000
1–3	5,000
4+	0

Table B.17
Termination Costs

Experience (years)	Percentage of Annual Salary
0–1	0
1–2	20
2–3	40
3–4	60
4–5	80
5+	100

Retirements

This model does not represent retirements.

Other Modeling Assumptions

In addition to the assumptions for the key variables outlined in the previous section, our model makes the following assumptions:

- The workforce is managed on a quarterly basis. In other words, decisions about growing or shrinking the workforce are made each quarter.
- Decisions to grow or shrink the workforce are made on the basis of the current quarter's demand, irrespective of future demand. In other words, the shipyard will try to hire as many workers as are needed to meet the current quarter's demand, and it will make redundant any workers that are unneeded to meet the current quarter's demand.
- Redundancy decisions are made in a way that preferentially favors retaining the most productive workers.
- Work that is uncompleted in a given quarter will propagate to the next quarter.
- Work is completed on a first-in, first-out basis. An implication of this is that work is not reprioritized once it enters the shipyard, and ships are always delivered in the planned order.
- A ship is not delivered until all the work in all skill categories is completed. An implication of this assumption is that delays in any individual skill categories will lead to delays in ship delivery.
- Workers can move between projects.

APPENDIX C
Sensitivity Analysis

In this appendix, we explore the sensitivity of our results to changes in key assumptions and variables. We focus primarily on assessing the influence of factors associated with shipbuilding demand because (1) these factors are the source of the most uncertainty, (2) the influence of these factors can be nuanced and insights can be gleaned from running and rerunning the simulation model, and (3) these factors can be directly influenced by the Australian government through acquisition and production plans. We also explore the effect of some workforce variables, such as hiring rate and proficiency, because these variables are a source of uncertainty even if they are largely outside the control of the government. Although there is uncertainty about the cost variables (particularly direct labor rates and overhead rates), the influence of these factors on cost outcomes is comparatively straightforward.

Table C.1 summarizes the main variables in our analysis and describes the base case and variants to be explored. See Appendix B for a more complete description of the variables and the assumptions.

Sensitivity Analysis of Short-Term Options for Sustaining a Fully Capable Shipbuilding Industrial Base

In this section, we examine how the variables affect the cost and schedule outcomes of different options for sustaining a fully capable shipbuilding industrial base (Path 1). As in Chapter Four, we assume a fully capable shipbuilding industrial base and a single "uber" shipyard. For each excursion, we examine the effect of the variables on two met-

Table C.1
Base Case Assumptions for Sensitivity Analysis of Workforce Sustainment Levels

Variables	Factor	Base Case	Variants Explored
Shipbuilding demand	Acquisition scenario	3 AWDs and 8 Future Frigates with planned delivery starting in 2026 and replacing the *Anzac* class	Variants discussed in Chapter Four
	Production plan	Future Frigates begin construction in 2020; 1-year drumbeat starting with Hull 3	Start date: 2017, 2018, 4th quarter 2018, and 2020 Drumbeat: 1 per year, 1 per 1.5 years, 1 per 2 years
	Demand profiles	Full build: 5 million man-hours, 22 quarters	1.8 million man-hours, 16 quarters 3 million man-hours, 20 quarters 7 million man-hours, 24 quarters
		Outfitting only: 2.5 million man-hours, 18 quarters	750,000 man-hours, 11 quarters 1.5 million man-hours, 14 quarters 3.5 million man-hours, 22 quarters
	Unit learning curve	5%	5% and 10%
Shipbuilding workforce	Initial labor	Meets 1st-quarter 2014 demand	N/A
	Labor pool	Sufficient to meet flux in remaining AWD program; does not replenish	N/A
	Hiring rate	10% (per quarter per skill category)	10%, 20%, 30% (per quarter per skill category)
	Firing rate	100%	N/A
	Workforce floor	Independent variable in workforce floor analysis	N/A
	Workforce ceiling	100% of peak future demand, by skill category	100% and 120% of peak future demand, by skill category
	Productivity	Per Table B.11	Per Table B.11 Accelerated: Twice as fast as Table B.10
	New hire distribution	Per Table B.12	N/A
	Mentoring ratio	4 to 1	N/A

Sensitivity Analysis

Table C.1—Continued

Variables	Factor	Base Case	Variants Explored
Shipbuilding cost	Direct labor rate	Per Table B.13	N/A
	Overhead rate	Per Table B.14 and B.15	N/A
	Training costs	Per Table B.16	N/A
	Termination costs	Per Table B.17	N/A
	Hiring costs	Assumed to be included in overhead rate	N/A
	Maximum overtime	10% (per skill category)	N/A

rics: (1) total labor cost and (2) total delay in the delivery of the Future Frigates, measured with respect to the planned retirement dates of the *Anzac*-class ships that they are intended to replace.

In short, we present the same basic data that are summarized in Table 4.3, but we present them graphically to facilitate a deeper understanding of sensitivities.

Level of Effort

Figure C.1 depicts the effect on cost of changing the total level of effort—whether the total level of effort required to produce a single ship is 1.8 million man-hours, 3 million man-hours, 5 million man-hours (the baseline), or 7 million man-hours.[1] Moving left to right across the figure, each column corresponds to a different level of effort, and the vertical axis represents the total labor cost. The black line shows the results for the base case (eight Future Frigates starting production in 2020), and each point corresponds to a different alternative scenario or plan as reflected in the legend. Figures C.2–C.24 follow this same structure for different variables.

The data confirm the obvious that larger ships would cost more, across all options. In a separate analysis not presented here, the cost per

---

[1] As discussed in Appendix B, our assumption is that the first-of-class ship requires 10 percent more additional effort and that follow-on ships follow a 95-percent unit learning curve. Notwithstanding these effects, for shorthand, we refer to different levels of effort by the central planning figure of 1.8 million, 3 million, 5 million, or 7 million man-hours.

Figure C.1
Effect of Level of Effort on Total Labor Cost (Full Capability Path)

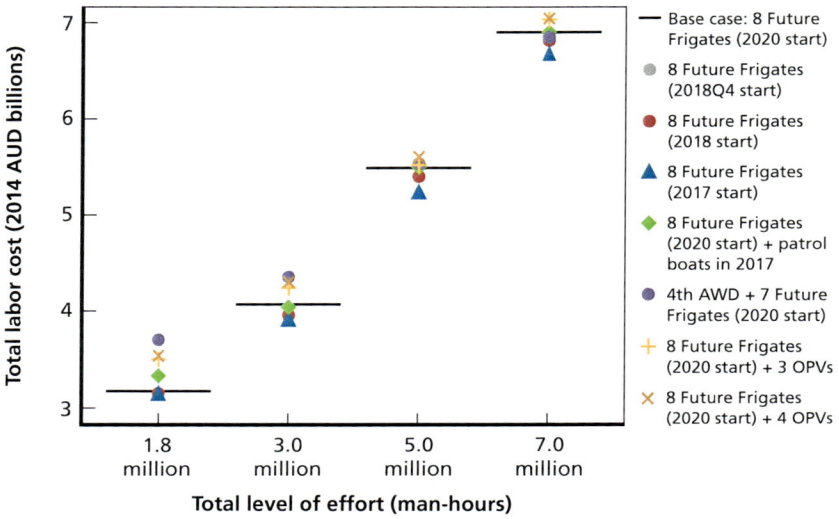

FTE worker decreases slightly, because the greater peak demand entails lower overhead rates on a per-FTE basis. But such efficiencies are not enough to outweigh the significant differences in the scale of the efforts.

The effect of level of effort on the relative attractiveness of the production plans is more subtle. In all cases, starting production early remains the cheapest option. But the relative attractiveness of adding a fourth AWD increases with increased level of effort. This is explained by the fact that the fourth AWD substitutes for the first Future Frigate; in such a one-for-one trade, the AWD appears more cost-effective the greater the effort (and thus more cost) required to produce the frigates. The relative attractiveness of adding OPVs or patrol boats has similar effects, although to a lesser degree for patrol boats, given their smaller size.

The effect on schedule relative to planned *Anzac* retirements also follows mostly predictable patterns; the results are depicted in Figure C.2. In general, larger ships require more time, and thus greater delays are expected for larger ships. There is no substitute for starting production

Figure C.2
Effect of Level of Effort on Total Schedule Delay (Full Capability Path)

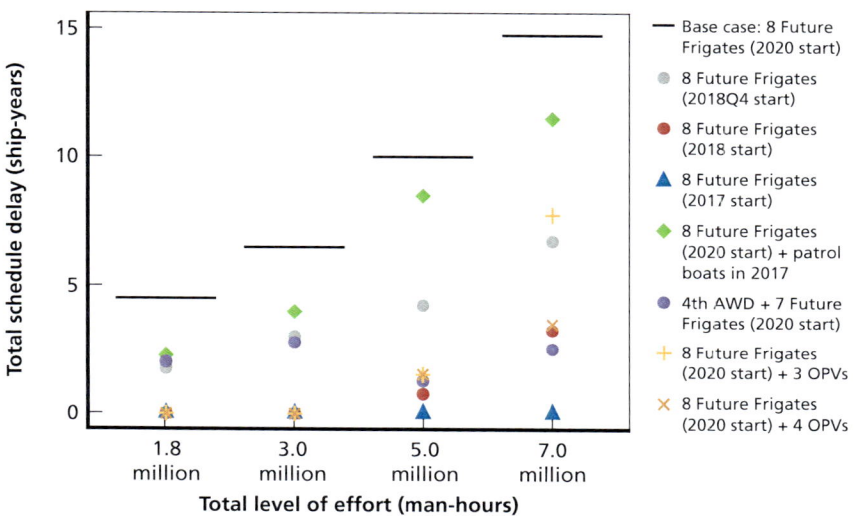

early, but adding a fourth AWD or OPVs can mitigate effects on schedule if starting the Future Frigates in 2020 is the chosen course.

However, again, a more subtle point can be made about the effect of adding a fourth AWD, which shows up as a purple marker in the figures. A separate analysis shows that a large part of the total delay is manifested by the fourth AWD itself, which would start production in 2017 after the shipyards have already shed workers due to declining workforce demands from the existing program. As the level of effort increases, the long-term peak demand increases, meaning that (under our assumptions) the shipyards would have a rational basis to raise the workforce ceiling in preparation for the coming surge. This means that for higher levels of effort, the capacity of the workforce could grow to higher levels during the production of the fourth AWD, reducing delay. Of course, there is a countervailing effect that larger ships simply take longer to build. These two effects compound to produce a "U" curve, where adding a fourth AWD yields the lowest delays among the options considered when the Future Frigates consume about 5 million fully productive man-hours.

Drumbeat

Figure C.3 depicts the effect on cost of changing the drumbeat—whether production of Future Frigate Hulls 4–8 are spaced by one year per hull (the base case), 1.5 years per hull, or two years per hull; all other variables are held at their baseline values. The data suggest a mostly predictable trend on the effect of drumbeats. Longer drumbeats lessen the peak demand, meaning that the shipyards do not have to grow to such high levels and then contract. On the other hand, higher peak demand from shorter drumbeats lessens overhead rates and creates smallish peaks and valleys, providing a countervailing effect on cost. The data suggest that a 1.5-year drumbeat balances these two effects, although the absolute value of the differences is rather small. The relative attractiveness of the options is essentially unaffected by drumbeat from a cost perspective.

Figure C.4 shows the effect on schedule of changing the drumbeat. The data show that, as expected, longer drumbeats lead to greater delays, simply because the start of production for the later frigates is pushed back. The relative attractiveness of the options is essentially unchanged from a schedule perspective.

Figure C.3
Effect of Drumbeat on Total Labor Cost (Full Capability Path)

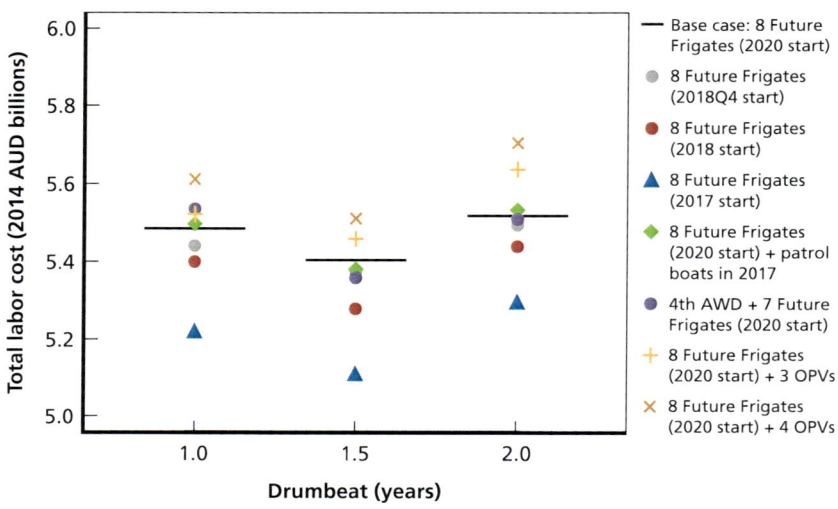

RAND *RR1093-C.3*

Figure C.4
Effect of Drumbeat on Total Schedule Delay (Full Capability Path)

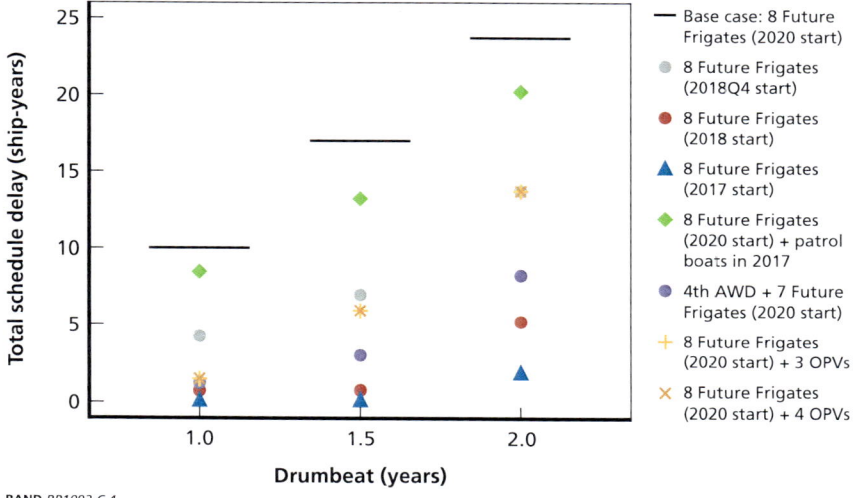

Unit Learning Curve

Figures C.5 and C.6 show the effects on cost and schedule, respectively, of unit learning curves—whether the Future Frigates are produced with a 95-percent or 90-percent unit learning rate. The data suggest that unit learning curve has only a limited effect on cost and essentially no effect on schedule outcomes; the relative attractiveness of the options is unchanged. One explanation for this is that the low purchase quantity of eight frigates does not give time for the effect of low learning rates to manifest in a significant way.

Hiring Rate

Figures C.7 and C.8 show the effects on cost and schedule of hiring rate—whether new workers can be hired at a rate of 10 percent per quarter of the current workforce per skill category (base case), 20 percent per quarter per skill category, or 30 percent per quarter per skill category. The data indicate that hiring rate could significantly affect schedule outcomes but is not likely to affect cost. One explanation of this is that a higher hiring rate means that the workforce can grow

Figure C.5
Effect of Learning on Total Labor Cost (Full Capability Path)

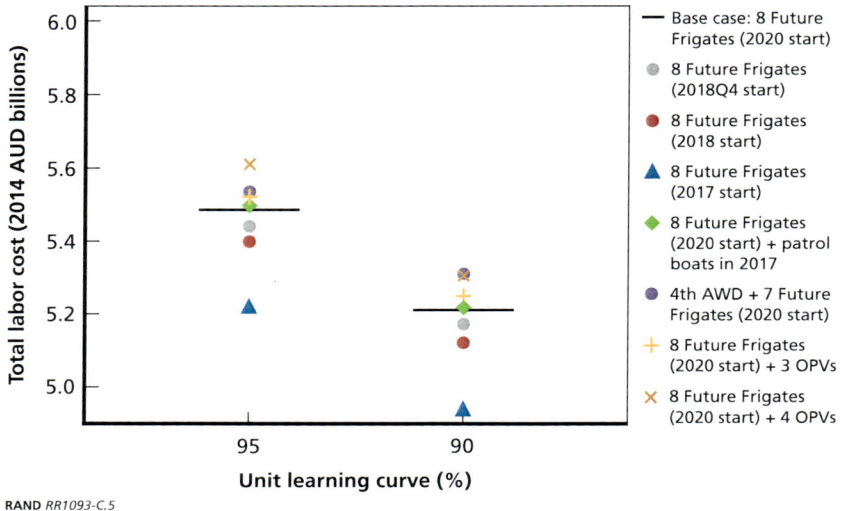

Figure C.6
Effect of Learning on Total Schedule Delay (Full Capability Path)

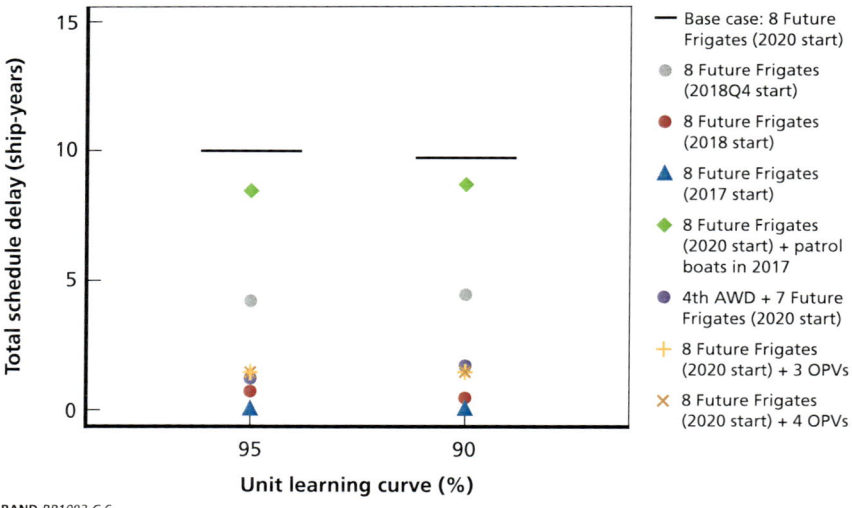

Sensitivity Analysis 199

Figure C.7
Effect of Hiring Rate on Total Labor Cost (Full Capability Path)

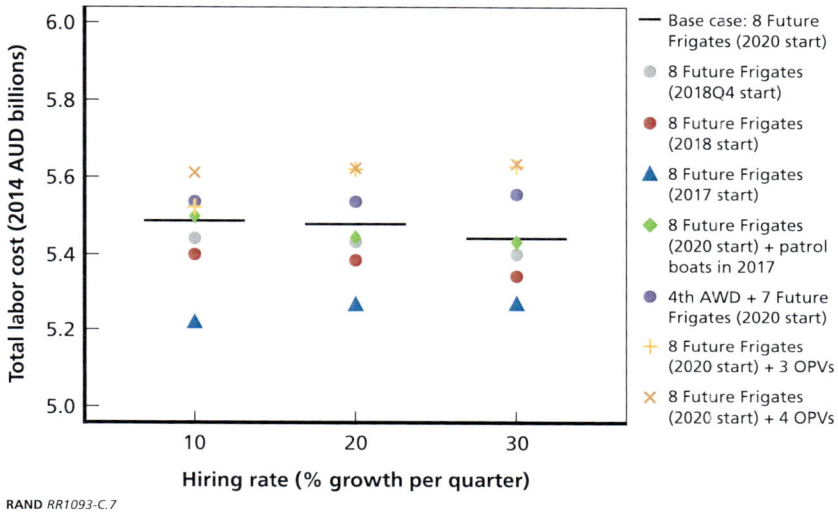

RAND RR1093-C.7

Figure C.8
Effect of Hiring Rate on Total Schedule Delay (Full Capability Path)

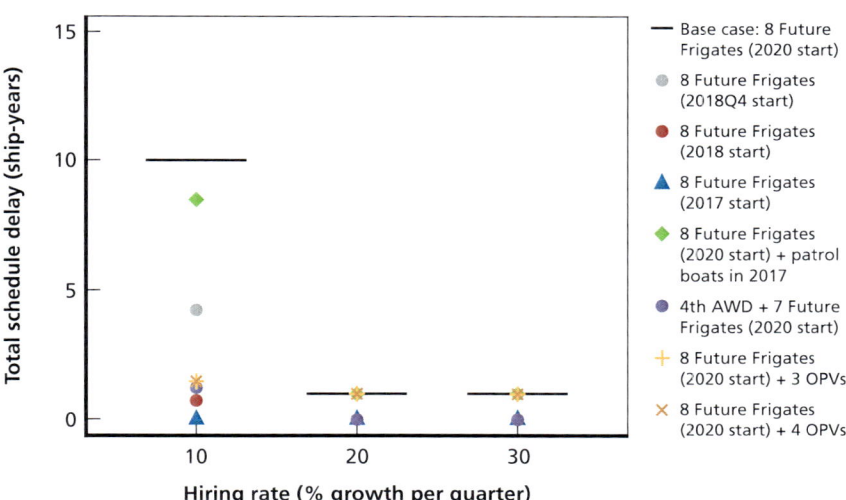

RAND RR1093-C.8

more quickly, but hiring rate does not change the number of workers that must be hired after a gap. Unfortunately, hiring rate is largely outside the control of the Australian government, so this reflects an assessment of uncertainty, not a policy lever. It may be prudent to plan for modest hiring rates.

Workforce Ceiling

Figures C.9 and C.10 show the effects on cost and schedule of workforce ceiling. As discussed in Appendix B, workforce ceiling reflects a shipyard policy of how many workers it will hire; it does not reflect the capacity of the labor market. The workforce ceiling comes into play when the shipyard falls behind on planned work; the assumption is that the shipyard would *not* hire unlimited new workers just to catch up. Rather, the shipyard would hire, at most, limited numbers in excess of peak planned demand to avoid the costs of hiring and terminating workers needed only temporarily.

The data suggest that workforce ceiling has only minimal effects on absolute values and no effect on the relative attractiveness of different production plans. A slight increase in cost results from a higher ceiling because shipyards with the flexibility to grow to larger workforces have higher training costs in growth periods and higher termination costs in periods of decline. This slight increase in cost comes with an advantage of slightly reduced delays, because a larger ceiling provides greater capacity to get work accomplished.

Productivity

Figures C.11 and C.12 show the effects on cost and schedule of worker productivity—the rate at which new hires gain proficiency. The data show that this variable can reduce costs and hasten deliveries in an absolute sense, because higher worker proficiency means that workforce gains productivity more quickly while it is regrowing, meaning more work is accomplished per FTE worker. However, this variable does not appreciably affect the relative attractiveness of the options.

Sensitivity Analysis 201

Figure C.9
Effect of Workforce Ceiling on Total Labor Cost (Full Capability Path)

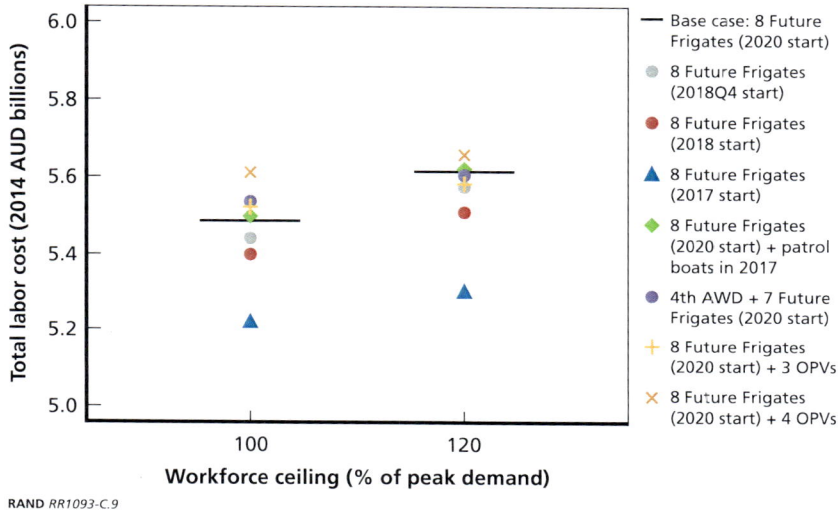

Figure C.10
Effect of Workforce Ceiling on Total Schedule Delay (Full Capability Path)

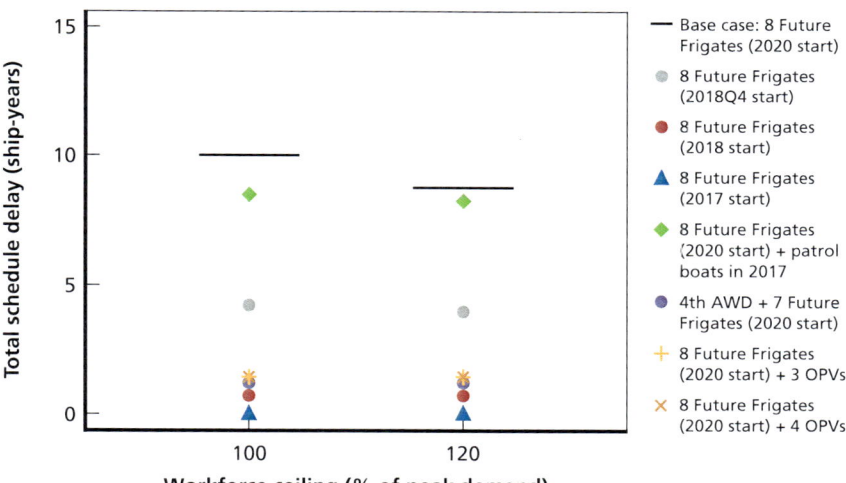

Figure C.11
Effect of Productivity on Total Labor Cost (Full Capability Path)

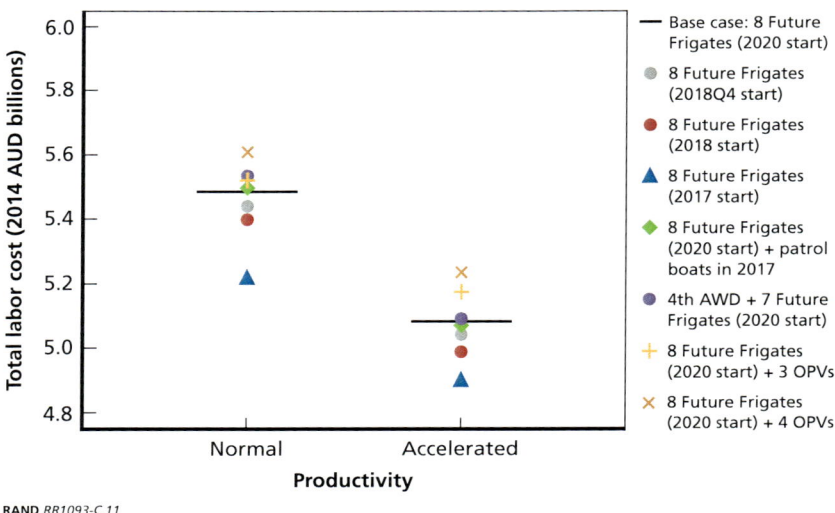

Figure C.12
Effect of Productivity on Total Schedule Delay (Full Capability Path)

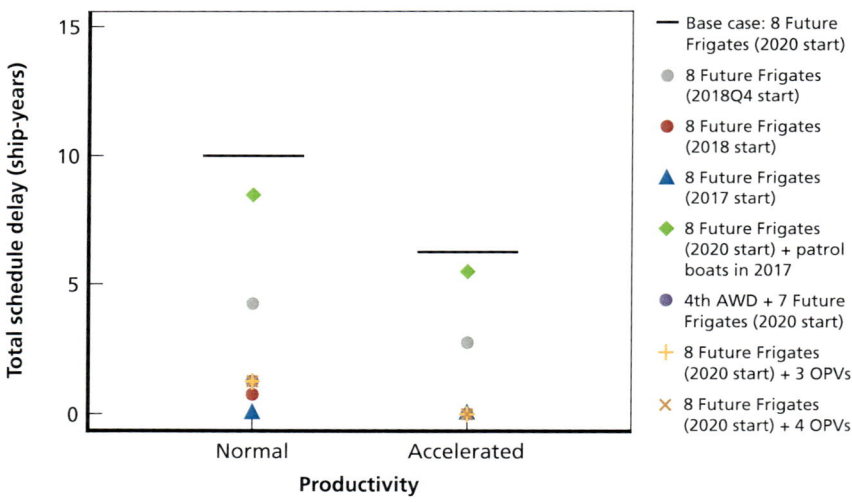

Conclusions

Several broad observations can be made from this sensitivity analysis. In general, the demand variables follow predictable trends. Larger levels of effort will be more costly and risk additional delays; longer drumbeats can increase delays and may save money by reducing peak demand so long as they are not too long (in this analysis, a 1.5-year drumbeat struck a balance between one- and two-year drumbeats, although the absolute differences are small); and unit learning rate has no significant effect, given the relatively small quantity of ships produced.

What is more interesting is how the relative attractiveness of options remains largely unchanged by these variables. First, regardless of level of effort, drumbeat, or unit learning curve, the best option from both a cost and schedule perspective is to start production of the Future Frigates early. In general, starting production in 2018 saves money and increases the chance of delivering ships in time to replace the retiring *Anzac* class; the effect on schedule is largely from having two additional years before the first retirement. There can be a notable difference even between starting production in 2017 and starting production in 2018, although as noted elsewhere in this document, it is likely impractical to start production before 2018 given the considerable design and contracting work that remains to be done in the Future Frigate program. If production cannot be started until 2018 or later, it appears that steps will be needed to mitigate cost and schedule implications of a production gap.

Second, in most cases, the option of adding a fourth AWD increases the overall productivity of the workforce and mitigates delays in delivery, but it increases total labor costs. Our analysis suggests that adding a fourth AWD *could* be cost-competitive from a labor perspective if the Future Frigate is the largest of the variants explored here (7 million man-hours). But even in this case, the cost savings would apply only if the fourth AWD replaces one of the eight frigates. Moreover, these costs do not include the significant costs to procure and integrate the Aegis combat system, nor to support such a large shape. As a general conclusion from examining the many cases presented here, adding a fourth AWD can mitigate risks to schedule, but the specific effects will depend on the level of effort.

Third, building the patrol boats in the major shipyards can improve productivity, and if the patrol boats start in 2017, there are very modest savings compared with the base case.[2] However, this option does not fully mitigate the effect of delayed delivery without also starting production of the frigates by 2018. Steps may be needed to ensure that the patrol boats themselves are not delayed after the start of the frigates.

Fourth, across the cases we examined, adding three or four OPVs during the production gap emerges as the most attractive option, assuming that it would be imprudent to start production of the Future Frigates early.

A range of other factors—such as hiring rate, workforce ceiling, and productivity—could well affect cost and schedule outcomes. However, variables such as these are not within the control of government decisionmakers, and thus represent an uncertainty, not a policy lever. Therefore, in general, it may be wise for government *not* to predicate acquisition and production plans on these variables.

Sensitivity Analysis of Short-Term Options for Sustaining a Limited Capability Shipbuilding Industrial Base

In this section, we examine how variables affect the cost and schedule outcomes of different options for sustaining a limited capability path (Path 2). For each excursion, we examine the effect of the variables on two metrics: (1) total labor cost and (2) total schedule delay in delivering the Future Frigates. In short, we present the same basic data that are summarized in Table 4.6, but we present them graphically to facilitate understanding. The analysis mirrors the preceding section on the full capability path (Path 1), and the results are presented in Figures C.13–C.24.[3]

---

[2] These savings come from building the patrol boats in the same shipyards where the Future Frigates will be built versus in other Australian shipyards.

[3] As in Chapter Four, we do not present analysis for adding OPVs in a limited capability path, because (1) the longer gap means that several more OPVs would be required to sustain the workforce in the gap, and (2) OPVs would sustain a much larger structural workforce than would be needed for an outfitting-only industrial base.

Sensitivity Analysis 205

Figure C.13
Effect of Level of Effort on Total Labor Cost (Limited Capability Path)

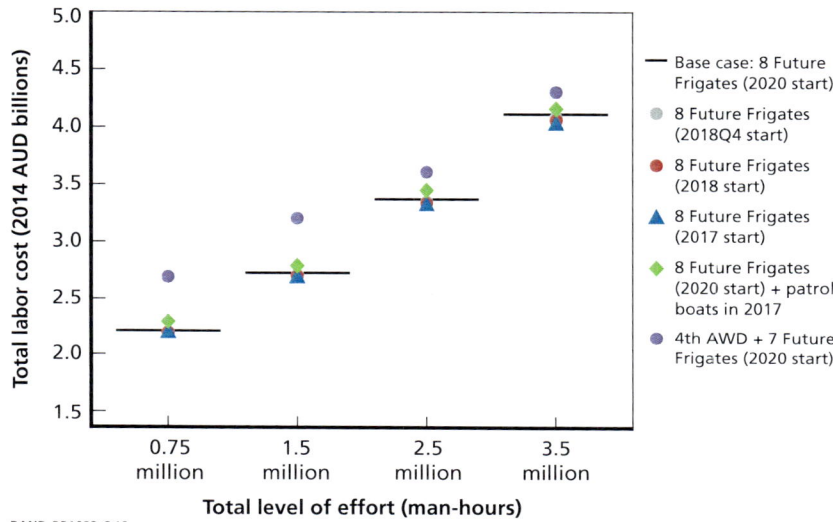

RAND RR1093-C.13

Figure C.14
Effect of Level of Effort on Total Schedule Delay (Limited Capability Path)

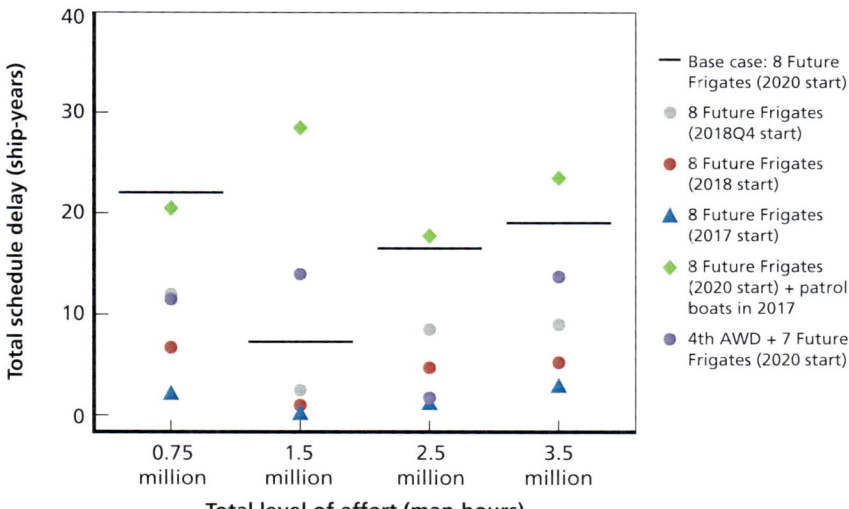

RAND RR1093-C.14

Figure C.15
Effect of Drumbeat on Total Labor Cost (Limited Capability Path)

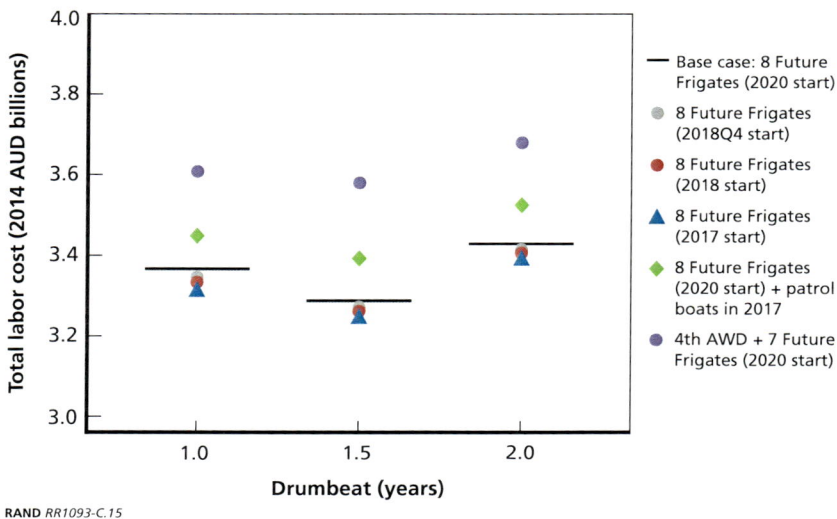

RAND RR1093-C.15

Figure C.16
Effect of Drumbeat on Total Schedule Delay (Limited Capability Path)

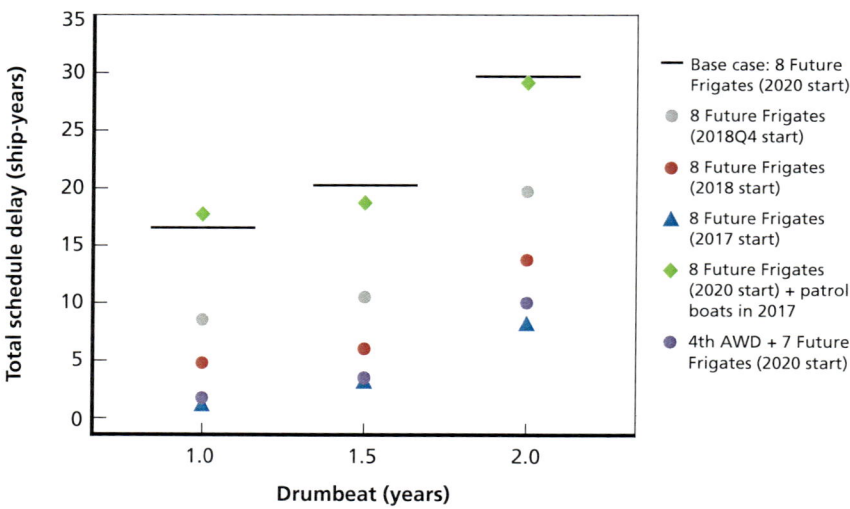

RAND RR1093-C.16

Sensitivity Analysis   207

Figure C.17
Effect of Learning on Total Labor Cost (Limited Capability Path)

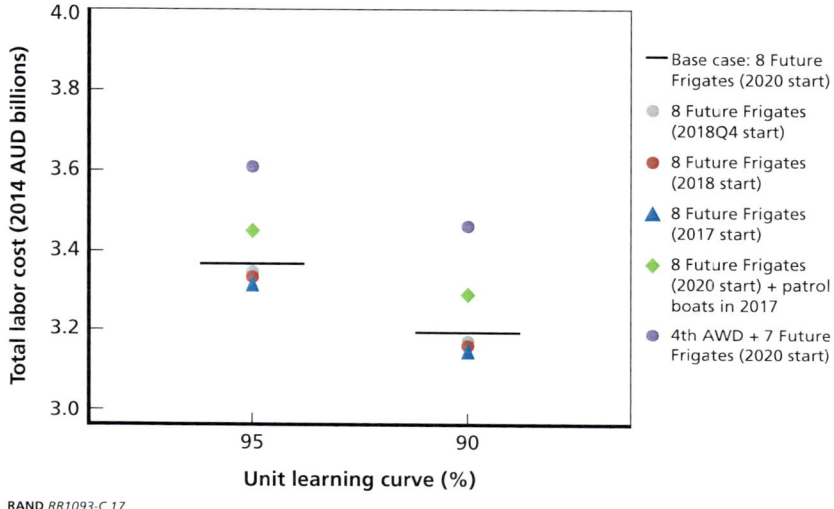

Figure C.18
Effect of Learning on Total Schedule Delay (Limited Capability Path)

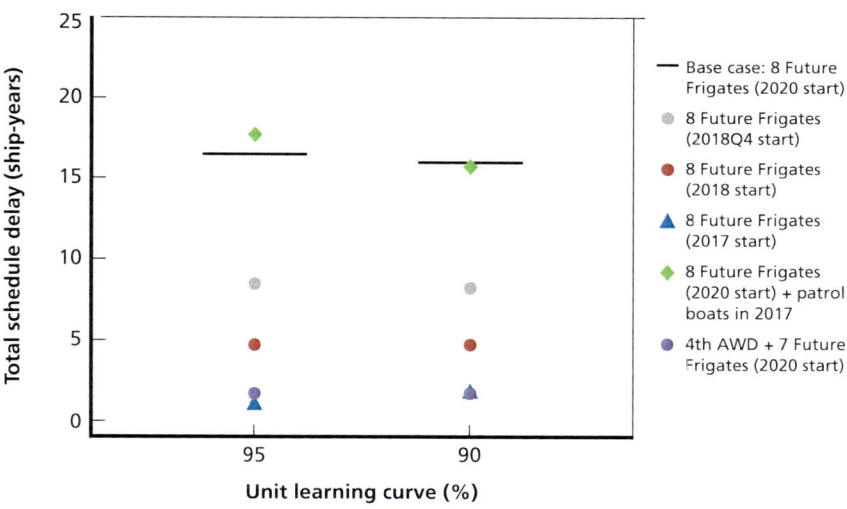

Figure C.19
Effect of Hiring Rate on Total Labor Cost (Limited Capability Path)

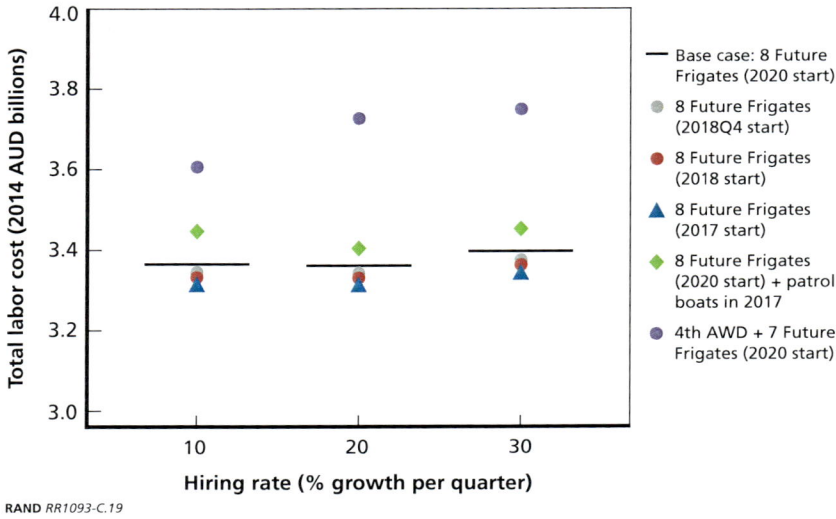

Figure C.20
Effect of Hiring Rate on Total Schedule Delay (Limited Capability Path)

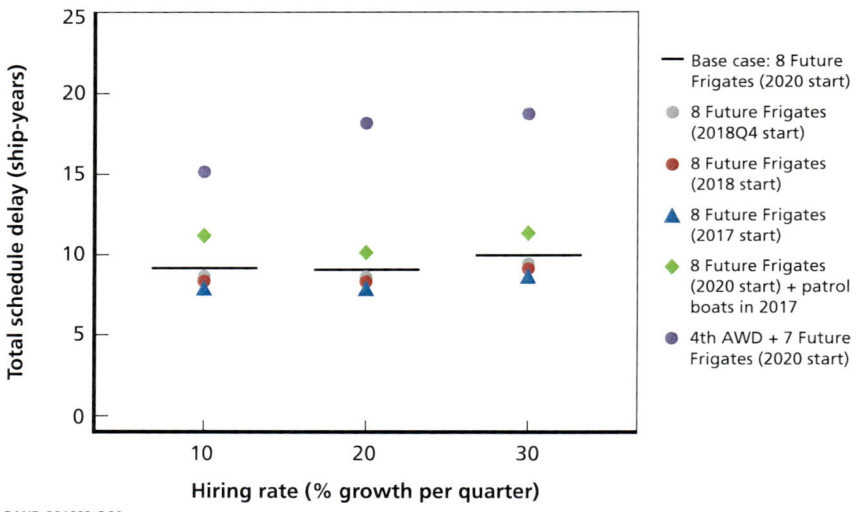

Sensitivity Analysis 209

**Figure C.21
Effect of Workforce Ceiling on Total Labor Cost (Limited Capability Path)**

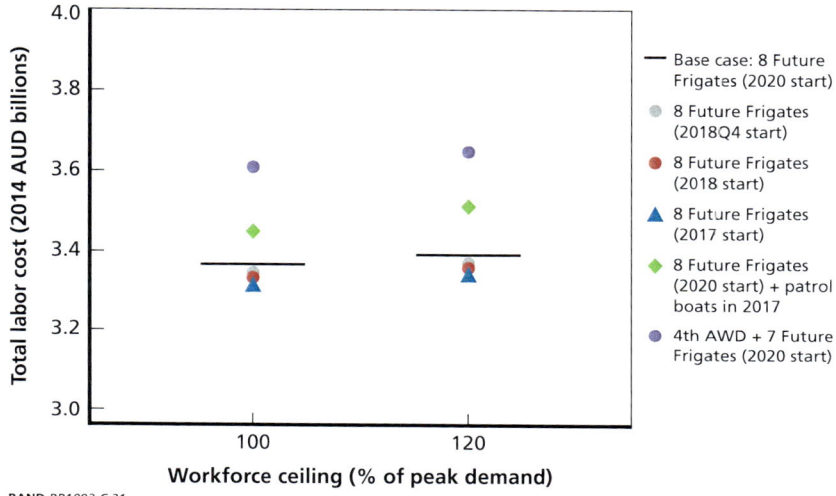

RAND RR1093-C.21

**Figure C.22
Effect of Workforce Ceiling on Total Schedule Delay (Limited Capability Path)**

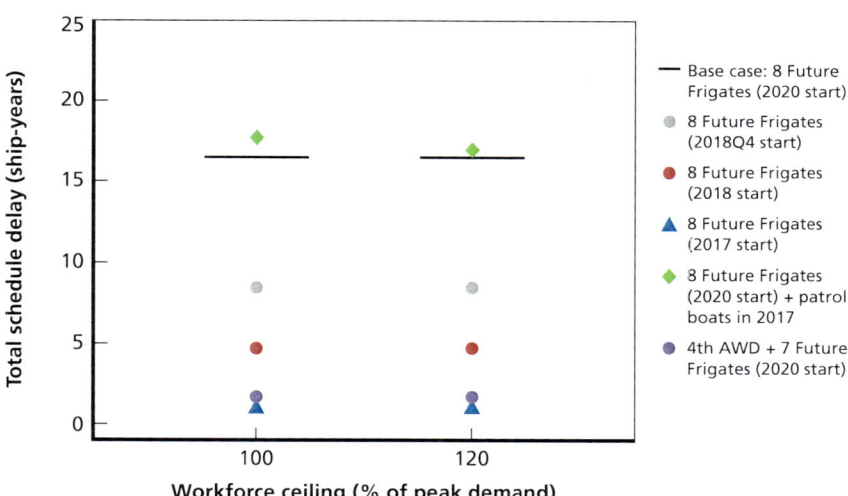

RAND RR1093-C.22

210 Australia's Naval Shipbuilding Enterprise: Preparing for the 21st Century

Figure C.23
Effect of Productivity on Total Labor Cost (Limited Capability Path)

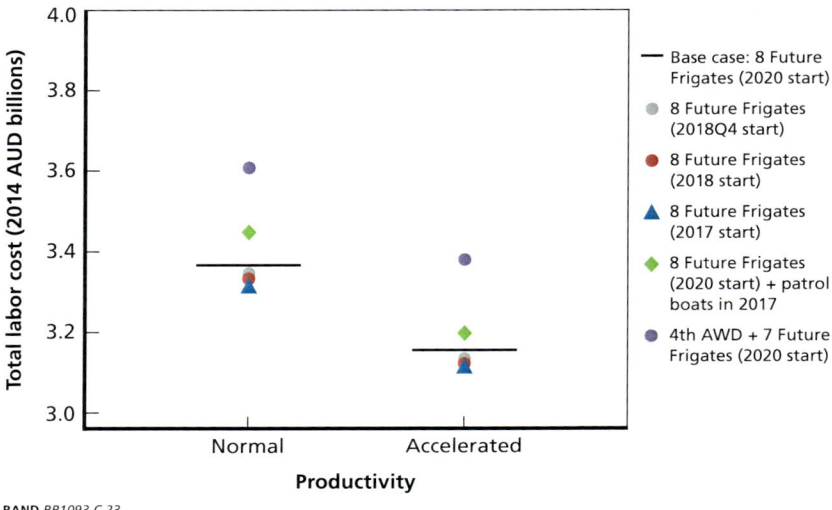

Figure C.24
Effect of Productivity on Total Schedule Delay (Limited Capability Path)

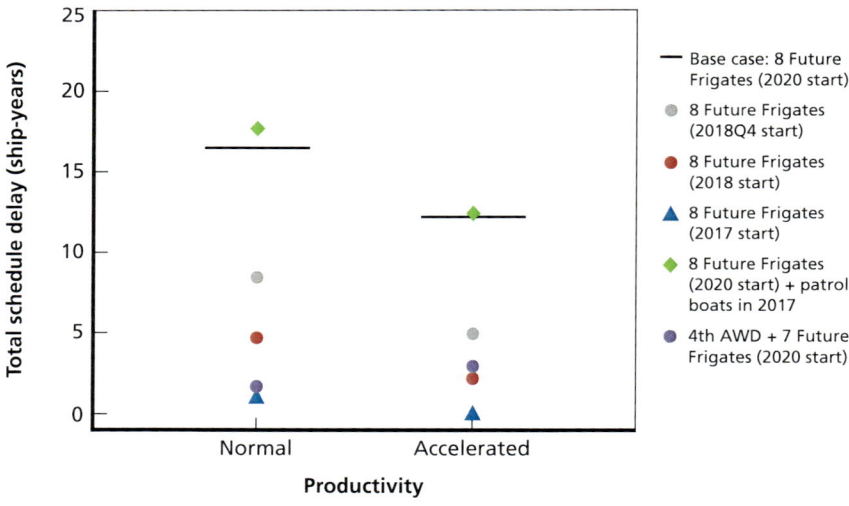

Conclusions

Interestingly, the conclusions from this analysis are consistent with that of the full-build case. The variables behave as expected and the relative attractiveness of the options is largely unaffected.

The one exception is the effect of total level of effort. As shown in Figure C.14, the total delay actually decreases with increased level of effort, from 0.75 million man-hours through 1.5 million man-hours. It then increases intuitively when the effort increases to 2.5 and 3.5 million man-hours. In general, increasing the total effort should increase delays because of the additional work required. However, in these cases, increased total effort also translates to increased peak demand, which increases the workforce ceiling by providing a rationale for the shipyard to hire more workers when work surges. In turn, the workforce has a greater capacity to recover from initial shortages after the gaps, and it sustains that capacity through the program. Thus, the greater total effort translates to greater ceiling, which translates to a greater capacity. The simple effect of additional work has a countervailing effect at 2.5 million man-hours.

APPENDIX D

Exploring the Option of Producing Offshore Patrol Vessels

The analysis presented in Chapter Four suggested that starting production of three to five OPVs in the 2017 time frame was an effective way of sustaining the shipbuilding workforce in the demand gap between the completion of the AWD program and the start of the Future Frigate. Our modeling of the shipbuilding workforce indicated that producing OPVs had only marginal increases in total labor costs (i.e., beyond the cost of producing eight Future Frigates), because the cost of the additional labor was largely offset by the cost savings associated with sustaining a productive workforce. Moreover, sustaining a productive workforce had the benefit of mitigating most of the delays in delivering Future Frigates to replace *Anzac*-class ships—delays that our modeling predicted would result if Australia needed to fully rebuild the workforce. These findings prompted AUS DoD to seek a deeper investigation into the potential implications of adding OPVs to the production plan. This appendix documents the additional analysis.

Scenarios and Assumptions

Let us first discuss the assumptions underlying this analysis and how those assumptions compare with the ones made in Chapter Four.

Industrial Structures and Shipbuilding Workforce

Based on analysis presented in Chapter Four, we focus on a fully capable shipbuilding industrial base (Path 1) because of our prior finding that an option of producing OPVs is not cost-effective for a limited capability industrial base (Path 2).[1] As in Chapter Four, we assume a single "uber" Australian shipyard, and we treat the decision of how to divide the work among multiple yards as a separate question. All variables relating to the shipyard workforce (e.g., hiring rates, attrition rates) and cost (e.g., direct labor and overhead rates, training and termination costs) are configured to their baseline values as documented in Appendix B.

Air Warfare Destroyer and Future Frigate Programs
This analysis focuses primarily on the base case assumptions regarding the AWD and Future Frigate programs. These are detailed in Appendix B and summarized in Chapter Four. To summarize briefly, we assume that the demands on the shipbuilding workforce to produce the AWDs follow the best available projections and that production of the eight Future Frigates starts in 2020. We assume that the Future Frigates nominally require 5 million man-hours and follow a 95-percent unit learning curve. We assume that there is a three-year gap between the start of the first of class and the second hull, and a two-year gap between the start of the second and third hulls. We examine two cases in which the remaining Future Frigates are produced on a one-year or two-year drumbeat. We refer the reader to Appendix B for a more detailed summary of these baseline assumptions.

Offshore Patrol Vessels
Our analysis focuses on several alternative plans for the production of OPVs. We examine two single-ship demand profiles that respectively assume OPV production nominally requires 700,000 man-hours or 500,000 man-hours. Independent of the total level of effort, we assume that the OPVs follow a 95-percent unit learning curve and that work is distributed across nine quarters. The 700,000 man-hour case is

---

[1] As discussed in Chapter Four, the longer production gap in Path 2 means that several more OPVs would be needed to sustain a workforce for the duration of the gap. More importantly, the production of OPVs (like patrol boats) would emphasize structural skills much more so than outfitting skills, when the latter is what requires sustaining in Path 2. As a result, adding OPVs would be a costly way to lessen the gap, and we do not examine it further in this report.

intended to approximate the effort to produce a 1,500-metric ton, 90-m vessel, whereas the 500,000 man-hour case represents a 1,000-metric ton, 70-m vessel. The single ship demand profile for the 700,000 man-hour case is shown in Figure D.1; the profile for the 500,000 man-hour case is simply a rescaled version of that profile.

We considered two cases: producing ten OPVs and 12 OPVs. The results for both were very comparable, so for present purposes, we discuss results only for the case of ten OPVs. We consider alternatives where OPV production starts in 2017, 2018, and 2019, and in all cases, OPVs are produced on a one-year drumbeat.

Differences Between Current and Baseline OPV Analysis

There are several noteworthy differences between the assumptions of the current analysis and the assumptions underlying the OPV analysis presented in Chapter Four. First, we examine a case with many more OPVs (ten to 12) than the original OPV analysis (three to five). Whereas the original case with three to five OPVs was intended purely to lessen

Figure D.1
Workforce Profile for One Offshore Patrol Vessel (700,000 Man-Hours, Nine Quarters)

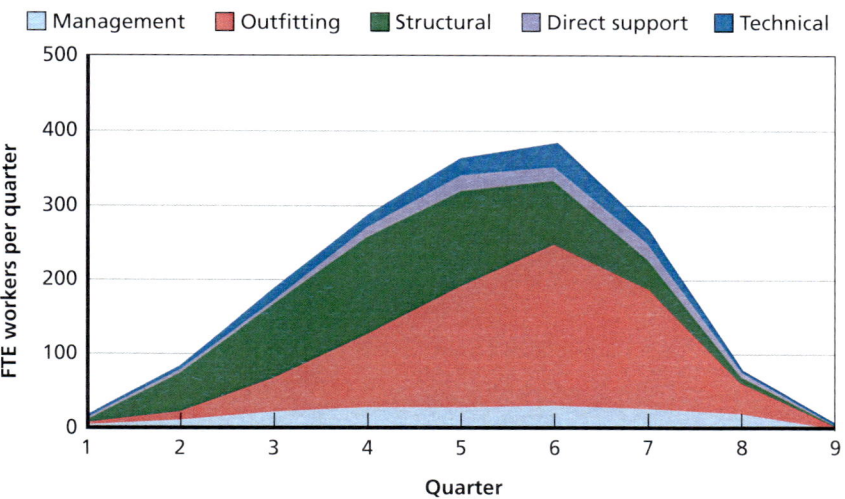

the workforce demand gap, a plan to produce ten to 12 OPVs reflects a potential operational requirement of the Australian Defence Forces.

Second, our baseline analysis assumes that the OPVs required 700,000 man-hours and 12 quarters to build, whereas in the current analysis, we assume that the builds are compressed into nine quarters, and we explore an additional variant requiring 500,000 man-hours. The shorter build duration reflects an attempt to accommodate the much larger procurement quantity in a comparable period of time. A nine-quarter build was deemed feasible by benchmarking the time required to produce similar vessels. The alternative level of effort is merely a kind of sensitivity analysis.

Third, the current analysis assumed that OPVs are produced on a regimented one-year drumbeat, again reflecting a notional operational requirement. The analysis in Chapter Four tailored the start of OPVs in an attempt to maximize the workforce sustained in the gap period. In practice, as we discuss in more depth below, if a primary objective of producing OPVs is to lessen the production gap, then shipyard managers could be strategic in deciding when hulls are started in order to level-load the shipyard workforce.

Results

As in Chapter Four, we assess the implications of different production plans using two metrics: the total labor cost and the total schedule delay in replacing the retiring *Anzac*-class frigates. Let us first consider the results under the assumption that the Future Frigates are produced on a one-year drumbeat, and then we will examine the case of a two-year drumbeat.

One-Year Drumbeat

Figure D.2 depicts the aggregate demand (in numbers of FTE workers) on the shipbuilding workforce under the assumption that the Future Frigates are produced on a one-year drumbeat and that the OPVs begin construction in 2019. For clarity of presentation, the plots discriminate

Figure D.2
Aggregate Workforce Profile for Building Offshore Patrol Vessels Starting in 2019, One-Year Drumbeat

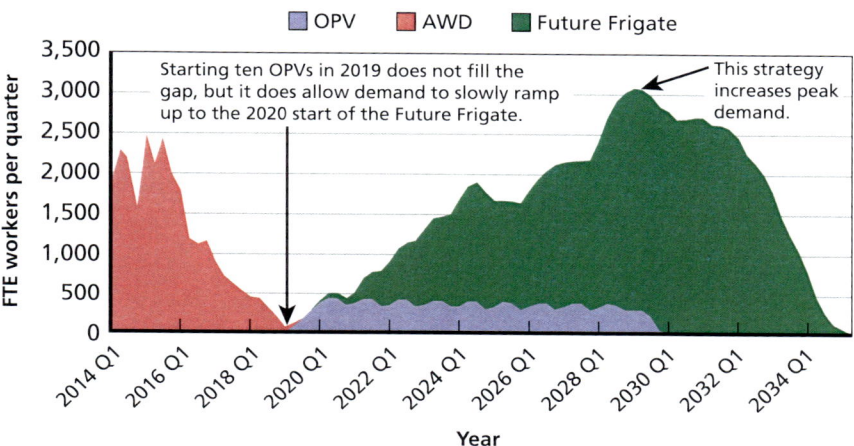

NOTE: The figure assumes that the OPVs require 700,000 man-hours to produce.
RAND RR1093-D.2

different programs rather than different skill categories (as opposed to what was presented in analogous charts in Chapter Four).

Several things can be observed from this chart. First, the demands associated with the AWD program are nearly if not completely finished by 2019, so starting the production of OPVs then would still require Australia to rebuild the workforce. Second, there is a more subtle effect of starting OPVs in 2019, which is that the ramp-up to start production of the Future Frigate is more gradual. As we shall see when we examine the modeling results, this has measurable impacts even if workforce is not fully sustained. Finally, the demands from the OPVs continue well into the late-2020s, when the Future Frigate program is in full swing. In particular, the peak demand on the shipbuilding workforce increases over the baseline, meaning that the workforce will have to rebuild to an even higher level.

Figure D.3 shows a similar demand profile under the assumption that the OPV construction begins in 2017. It shows that starting the OPVs in 2017 sustains the workforce across the gap, and although the

Figure D.3
Aggregate Workforce Profile for Building Offshore Patrol Vessels Starting in 2017, One-Year Drumbeat

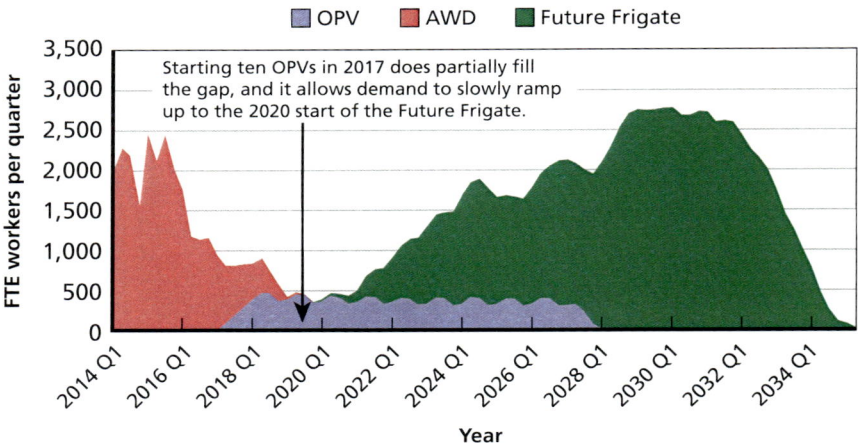

NOTE: The figure assumes that the OPVs require 700,000 man-hours to produce.
RAND *RR1093-D.3*

OPV production extends well into the Future Frigate program, the OPVs finish earlier and thus do not increase the peak demand.

The cost and schedule implications of these scenarios were examined using RAND's Shipbuilding and Force Structure Analysis Tool. Figures D.4 and D.5 show the results. The black bar shows the baseline case (i.e., without producing any OPVs), and the red and blue bars show results for the 700,000 man-hour and 500,000 man-hour OPVs, respectively. Results are provided for different cases corresponding to 2017, 2018, and 2019 starts for the production of the OPVs.

Several broad observations can be made from examining these results. First, in general, adding OPVs as planned only marginally increases the total labor costs. This is because most of the additional labor costs for producing OPVs are offset by the savings associated with sustaining a productive workforce or more gradually ramping up the Future Frigate program. Unsurprisingly, the increased costs are greater for the 700,000 man-hour OPVs than the 500,000 man-hour OPVs.

Exploring the Option of Producing Offshore Patrol Vessels   219

**Figure D.4
Total Labor Costs for Building Ten or 12 Offshore Patrol Vessels, One-Year Drumbeat**

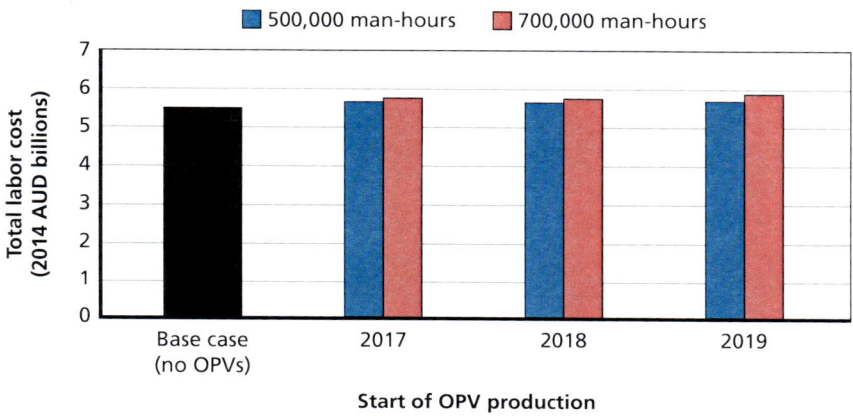

**Figure D.5
Total Schedule Delay for Building Ten or 12 Offshore Patrol Vessels, One-Year Drumbeat**

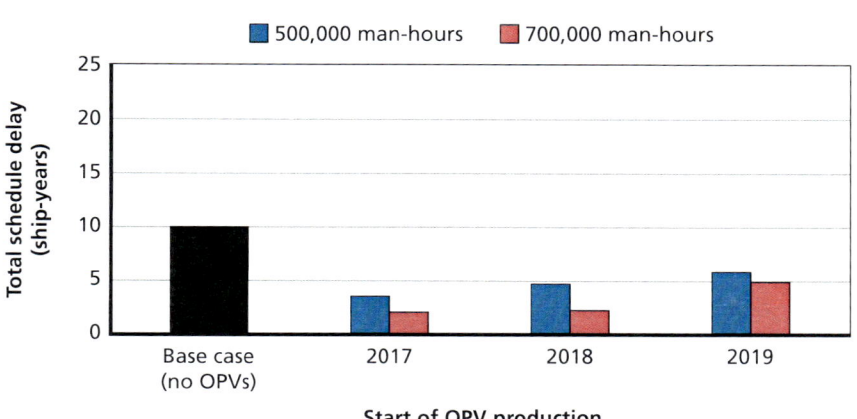

Second, in general, adding OPVs mitigates much of the delays associated with replacing the retiring *Anzac* class with Future Frigates that start production in 2020. For example, in the base case, our model predicted a delay of ten ship-years in replacing the *Anzac* class, but starting production of ten 700,000 man-hour OPVs in 2017 reduces that delay to less than two ship-years. The delays are modestly better for a 700,000 man-hour Future Frigate than a 500,000 man-hour one, because there is more work to sustain a larger productive workforce.

Third, the benefits of lessening the gap with OPVs dissipate the later OPV production starts. For example, starting production of the OPVs in 2019 increases the total delay to five ship-years. This is still less than the baseline of ten years but greater than starting OPVs in 2017. There may also be cost improvements associated with starting OPVs earlier, although the numeric differences in our results are smaller than the precision afforded by our model.

Two-Year Drumbeat

Figures D.6 through D.9 provide analogous depictions of the results under the assumption that Future Frigates are produced on a two-year drumbeat. As with the one-year drumbeat, starting production in 2019 does not lessen the gap but allows for a more gradual ramp up into the Future Frigate program. However, unlike the one-year drumbeat case, the addition of OPVs does increase the peak demand (over the base case of producing Future Frigates on a two-year drumbeat without OPVs) regardless of when the OPVs start. This is expected because the effect of the two-year drumbeat is to level-load the industrial base by distributing the work over time, meaning the peak demands for the Future Frigate program are reduced but occur earlier.

The conclusions about the efficacy of adding OPVs on the Future Frigate program are identical as for the one-year drumbeat. However, as noted in Chapter Four, there are delays inherent to a two-year drumbeat that cannot be mitigated by any gap-lessening strategy. The issue is that the *Anzac*-class ships were delivered on a one-year drumbeat, and with a 30-year lifespan, they will retire at that rate as well.

Exploring the Option of Producing Offshore Patrol Vessels   221

Figure D.6
Aggregate Workforce Profile for Building Offshore Patrol Vessels Starting in 2019, Two-Year Drumbeat

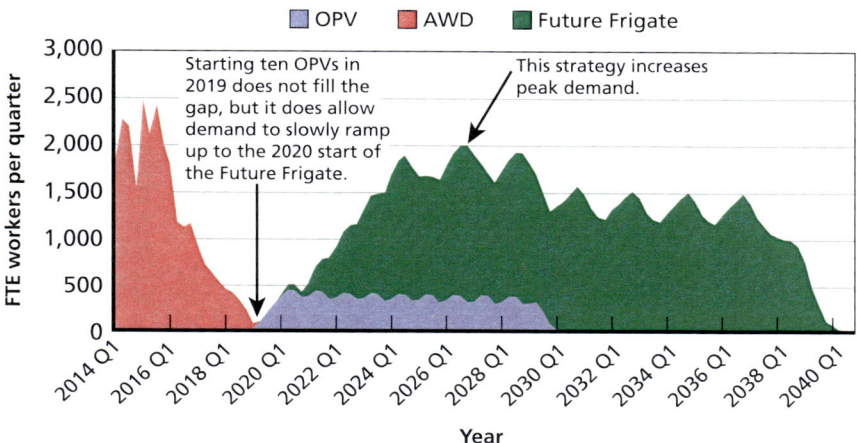

NOTE: The figure assumes that the OPVs require 700,000 man-hours to produce.
RAND RR1093-D.6

Figure D.7
Aggregate Workforce Profile for Building Offshore Patrol Vessels Starting in 2017, Two-Year Drumbeat

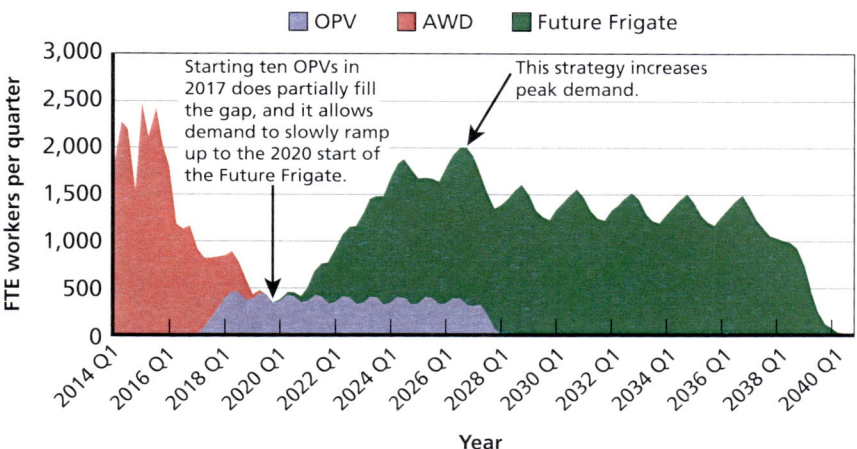

NOTE: The figure assumes that the OPVs require 700,000 man-hours to produce.
RAND RR1093-D.7

Figure D.8
Total Labor Costs for Building Ten or 12 Offshore Patrol Vessels, Two-Year Drumbeat

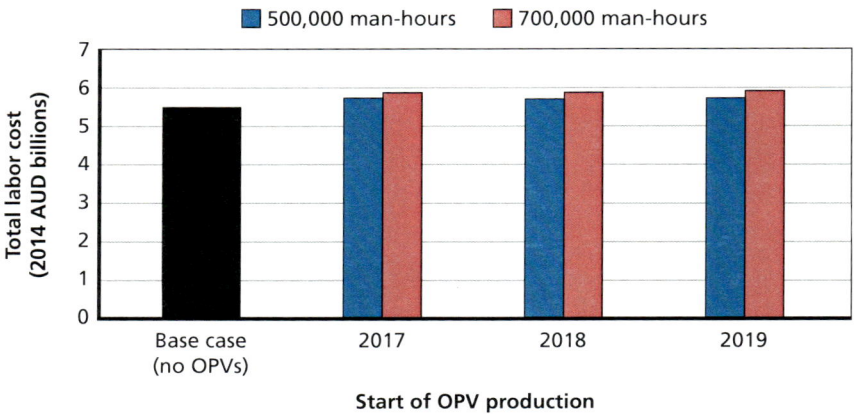

Figure D.9
Total Schedule Delay for Building Ten or 12 Offshore Patrol Vessels, Two-Year Drumbeat

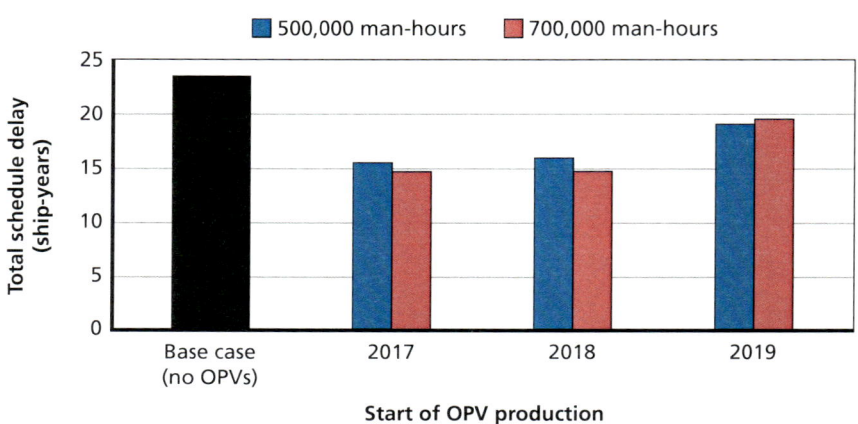

Conclusions

A production gap looms between 2017, when the AWD workforce demands begin to drop in significant measure, and 2022, when demands to produce the Future Frigates ramp up. Future Frigates are likely to cost more and be delivered late if Australia chooses a path to produce them indigenously but does not take steps to sustain a productive shipbuilding workforce during the gap. As discussed in Chapter Four, starting the Future Frigates early is one option, but doing so is technically infeasible and risky, given that requirements and designs have not been finalized.

Both the analyses in Chapter Four and in this appendix indicate that producing OPVs is an attractive option for sustaining a productive workforce during the gap. Producing OPVs does not add significantly to labor costs, yet it mitigates most of the schedule delays in replacing the retiring *Anzac*-class ships.

The specific effects of adding OPVs on Future Frigate cost and schedule depend on a number of factors that remain uncertain. Most importantly, they depend on the requirements and the build strategy for the Future Frigate, which will determine the size and duration of that program, as well as the number of shipyards that will be involved. But, of course, the effects also depend on the number of OPVs to be produced. Producing too few OPVs may be insufficient to lessen the gap, and producing too many OPVs may increase the peak demand and introduce competition for shipyard workers between OPVs and Future Frigates. The effects also depend on the size and duration of OPV builds; smaller or shorter builds may mean that more OPVs are needed to lessen the gap.

Our analysis also shows that the effect of building OPVs can be tuned through careful design of the OPV production plans. In Chapter Four, we showed that as few as three OPVs could lessen the gap if the builds were stretched out to 12 (versus nine) quarters and placed strategically within the gap period. The analysis presented in this appendix shows that as many as ten OPVs could have comparable effects.

Ultimately, if the primary objective of adding OPVs is sustaining the workforce to lessen the workforce demand gap, then program and shipyard managers can use OPV build strategy as a lever to optimize those effects.

Industrial Structure

This analysis assumed that ships were produced in a single "uber" shipyard. Of course, if Australia pursues an industrial strategy of supporting multiple shipyards, then the OPVs would need to be distributed across those yards. The specific details will depend on what work the second or third yards will do in the Future Frigate program. However, in general, adding more yards means that more OPVs will be required.

Guidelines for Using OPVs to Lessen the Gap

Notwithstanding these large unknowns, our analysis suggests several guidelines for using OPV production as a strategy for sustaining the shipbuilding workforce in the gap period.

- *Start production of the OPVs by 2017, or as soon as possible.* By 2017, demands from the AWD program will begin to fade, and it will be necessary to sustain the productivity of core workers before they find work in other industries.
- *Use an existing OPV design without modifications.* The risks of creating a new design, or Australianizing an existing design, include delaying the start of the OPVs or introducing production delays later in the OPV program. Australia can optimize the chance of sustaining a productive workforce if it uses an existing design without modifications.
- *Aim to sustain 20–30 percent of peak Future Frigate demands.* Our workforce demand analysis in Chapter Four shows that sustaining this level increases total costs only marginally, reduces unproductive labor, and mitigates the delays in delivering the Future Frigates.
- *Schedule successive OPV builds strategically.* This analysis made particular assumptions about the duration of OPV builds and the drumbeat for producing them. However, if the objective of building OPVs is to sustain a workforce, then these parameters can be chosen to meet this objective. More generally, the production

plans will need to balance the workforce needs for the start of the frigates; deliver OPVs on time to meet any operational requirements; and avoid having OPVs compete with Future Frigates for shipyard labor.
- *If there are multiple shipyards, distribute OPV work in a way that complements the distribution of Future Frigate work.* Once Australia decides on an industrial strategy, OPVs should be distributed across shipyards in a way that aligns with their role in the Future Frigate program.

Shipyard Capacity

A natural question is whether Australia's shipyards have the capacity to produce AWDs, Future Frigates, and OPVs. This analysis focused on the workforce and did not include a detailed analysis of facilities. However, some insights can be gleaned from a straightforward examination of the demand profiles.

For purposes of this discussion, let us assume that OPVs begin construction in 2017 to optimize the effects of sustaining a productive workforce. Under this assumption, the AWDs will be nearing delivery when the first OPV begins construction. The facilities involved with block construction of the OPVs would have minimal overlap with the facilities required for the final phases of AWD outfitting, so there is unlikely to be a bottleneck. By 2019, there could be multiple OPVs in a shipyard at a given time, assuming that OPVs are produced on a one-year drumbeat. However, today there are multiple AWDs in the yard simultaneously, so it is difficult to imagine that facilities would be stressed by the demands of multiple OPVs, a much smaller vessel than an AWD.

In the 2020–2023 time frame, the production of the first-of-class Future Frigates will have begun, and at that time, it is conceivable that block construction facilities will be in high demand. But again, noting Australia's recent experience having three AWDs in construction simultaneously, it is difficult to imagine insurmountable challenges with having even three OPVs and a Future Frigate being built simultaneously.

The more significant challenge may occur after 2023, when there are multiple Future Frigates in the yards at the same time that the OPVs are in production. Without a detailed analysis of facilities, it is difficult to assess whether there would be any bottlenecks, but they are plausible. Such constraints could be addressed, by either focusing more of OPV production in a second shipyard or building additional facilities.

APPENDIX E

Survey of Australian Shipbuilders and Ship Repair Industries

This appendix reproduces the survey used to collect information from organizations making up the Australian naval shipbuilding industrial base.

Australian Naval Shipbuilding Industrial Base Study

Introduction

The Australian DoD has asked RAND to conduct a study of the Australian Naval Shipbuilding Industrial Base examining the next two decades of production. This study is principally focused on issues related to workforce and employment for the various skills and trades. However, we do ask for information about facilities. Please feel free to attach or append any supporting information that you feel is appropriate.

Defence has also asked for copies of the data that your firm provides. You might consider some of the data to be proprietary or business sensitive. Also, we expect to have nondisclosure agreements in place. Therefore, we ask you to please indicate whether RAND may share the data with Defence.

- **Yes – RAND may share these data with Defence**
- **No – RAND may not share these data with Defence**

After we have received your completed survey form, we are willing to follow up through a site visit, phone conversation, and/or email. We will leave the choice up to you. Please let us know your preference in terms of follow-on questions and clarifications to the completed form.

Thank you for your assistance with this study.

Defence Contacts

Our Defence sponsor for this work is the White Paper Enterprise Management Team.

The contact is: Dr. Darren J. Sutton
White Paper Enterprise Management Team
Darren.sutton@defence.gov.au
(02) 6265 7816
0455 064 050

RAND Contacts

If you have any questions or require further clarification, please contact one of the RAND analysts listed below:

John Schank   John Birkler
RAND   RAND
1200 South Hayes St.   1776 Main Street
Arlington, VA 22202-5050   Santa Monica, CA 90407
1-703-413-1100, ext 5304   1-310-463-1924
schank@rand.org   birkler@rand.org

Persons Completing the Form

Name	Title	Phone #	Email Address

Shipyard Labor

Instructions

In this section, we request information concerning current workforce employment levels, demographics, training, proficiency, and so forth. The questions are focused on data that are, hopefully, readily available through your human resources department and/or long-range planning group. Most of the questions ask for values to be entered into a table. These tables are organized at two levels of detail: category and subcategory (see Table R.1 below). For example, the specific skills of engineering, design (including drafting and CAD personnel), estimating, planning, and program management/control comprise the subcategory of *technical*. The subcategories *technical* and *general management* comprise the category *management and technical*.

Table R.1
Categories, Subcategories, and Specific Skills for Labor

Category	Subcategory	Specific Skill
General management and technical	General management	Management
		Administration
		Marketing
		Purchasing
	Technical	Design
		Drafting/Computer Aided Design (CAD) specialist
		Engineering
		Estimating
		Planning
		Program control/project management
Manufacturing	Structure	Steelworker, plater, boilermaker
		Structure welder
		Shipwright/fitter
		Team leader, foreman, supervisor, progress control (fabrication)
	Outfitting	Electrician, electrical tech, calibrator, instrument tech
		Heating, ventilation, and cooling (HVAC) installer
		Hull insulator
		Joiner, carpenter
		Fiberglass laminator
		Machinist, mechanical fitter/tech, fitter, turner
		Painter, caulker
		Pipe welder
		Piping/machinery insulator
		Sheet metal
		Team leader, foreman, supervisor, progress control (outfitting)
		Weapon systems
	Direct support	Rigger, stager, slingers, crane, and lorry operators
		Service, support, cleaners, trade assistant, ancillary
		Stores, material control
		Quality assurance/control

1. Please provide the number of your company's employees (yearly average) over the past five years.

Category	Subcategory	2009	2010	2011	2012	2013
Management and technical	General management					
	Technical					
Manufacturing	Structure					
	Outfitting					
	Direct support					
Other (please specify)						

2. Please provide the number of workers for your current workforce for each of the three pairs: direct versus indirect employees, temporary versus permanent employees, and part-time versus full-time employees. A temporary employee is defined as someone who is directly employed by your company or a subcontractor for a specified/limited period of time. A part-time employee works less than a normal full-time schedule but does not have a specified term of employment.

Category	Subcategory	Indirect	Direct	Temporary	Permanent	Part Time	Full Time
Management and technical	General management						
	Technical						
Manufacturing	Structure						
	Outfitting						
	Direct support						
Other (please specify)							

3. Please provide your current workforce age distribution.

Category	Subcategory	< 21 Years Old	21 to 30 Years Old	31 to 40 Years Old	41 to 50 Years Old	51 to 60 Years Old	> 60 Years Old
Management and technical	General management						
	Technical						
Manufacturing	Structure						
	Outfitting						
	Direct support						
Other (please specify)							

4. Please provide the current distribution of your workforce by years of experience in the marine engineering sector (i.e., shipbuilding, ship repair, and offshore fabrication).

Category	Subcategory	< 1 Year	1 Year	2 Years	5 Years	10 Years	20 Years	> 20 Years
Management and technical	General management							
	Technical							
Manufacturing	Structure							
	Outfitting							
	Direct support							
Other (please specify)								

5. Please provide the number of annual recruits for the past five years.

Category	Subcategory	2009	2010	2011	2012	2013
Management and technical	General management					
	Technical					
Manufacturing	Structure					
	Outfitting					
	Direct support					
Other (please specify)						

6. Please indicate the typical experience level of your permanent, new hires as a percentage of those hired.

Category	Subcategory	< 1 Year	1 Year	2 Years	3 Years	4 Years	5 Years	> 5 Years
Management and technical	General management							
	Technical							
Manufacturing	Structure							
	Outfitting							
	Direct support							
Other (please specify)								

7. Please indicate the typical experience level of your temporary hires as a percentage of those hired.

Category	Subcategory	< 1 Year	1 Year	2 Years	3 Years	4 Years	5 Years	> 5 Years
Management and technical	General management							
	Technical							
Manufacturing	Structure							
	Outfitting							
	Direct support							
Other (please specify)								

8. Please provide the direct hourly wage rate for your employees in the past five years.

Category	Subcategory	2009	2010	2011	2012	2013
Management and technical	General management					
	Technical					
Manufacturing	Structure					
	Outfitting					
	Direct support					
Other (please specify)						

9. Please provide your annual training cost per worker by experience and subcategory.

Category	Subcategory	< 1 Year	1 Year	2 Years	3 Years	4 Years	5 Years	> 5 Years
Management and technical	General management							
	Technical							
Manufacturing	Structure							
	Outfitting							
	Direct support							
Other (please specify)								

10. Please indicate the relative productivity (percentage relative to the highest skilled worker) in naval shipbuilding by experience and skill category. For each subcategory, we have assumed that workers beyond five years of work experience (in a general area) are at the highest level of productivity, which we defined to be 100%. Thus, if a worker with two years of experience in the naval shipbuilding sector is half as productive as one with five years of experience, then you should enter 50% in the box for "2 years."

Category	Subcategory	< 1 Year	1 Year	2 Years	3 Years	4 Years	5 Years	> 5 Years
Management and technical	General management							
	Technical							
Manufacturing	Structure							
	Outfitting							
	Direct support							
Other (please specify)								

11. Please provide the average age of your workers at the time of their retirement. Does it vary by subcategory?

12. Please provide the number of losses in the past five years *not* due to lay-offs or redundancies (e.g., voluntary departures, retirements, long-term disabled, etc.).

Category	Subcategory	2009	2010	2011	2012	2013
Management and technical	General management					
	Technical					
Manufacturing	Structure					
	Outfitting					
	Direct support					
Other (please specify)						

13. Please provide the number of losses in the past five years due to lay-offs or redundancies.

Category	Subcategory	2009	2010	2011	2012	2013
Management and technical	General management					
	Technical					
Manufacturing	Structure					
	Outfitting					
	Direct support					
Other (please specify)						

14. Over the next several years, do you anticipate problems maintaining an adequately sized workforce? Please explain.

15. Are there particular worker skills that are in high demand or for which recruiting is difficult? Please explain.

16. How many annual work hours do you plan for your employees (not counting overtime)?

Current and Future Production Plans

We would now like to understand the projects you have completed in the past five years and any projects currently under way that will extend into the future. For example, a project may be "Build first AWD" or "perform major repair on XXX." We first ask for the list of projects and then request a separate sheet showing the quarterly demand (in number of man-hours) over the life of the project for the various subcategories. Projects that have been completed should reflect actual workloads, and projects that are still ongoing will reflect actual and projected workloads.

17. Please provide information concerning your past and current projects. *Note: If there are more than 15 activities planned, please expand the list.*

Project	Name/ Description	Type of Work (e.g., new, repair, module, etc.)	Start of Design and Planning (month/year)	End of Design and Planning (month/year)	Start of Production (month/year)	Delivery (month/year)
1						
2						
3						
4						
5						
6						
7						
8						
9						
10						
11						
12						
13						
14						
15						

18. For each project listed in question 17, please provide a separate sheet showing the workload in number of man-hours per quarter over the life of the project. A sample sheet is shown below and can be replicated as often as needed.

Activity
Name:_____
Type of work:_____

Number of Man-Hours Each Quarter

Please provide the total workload, by quarter, for the activities listed. The first quarter is the start of work on the ship (regardless of calendar date):

Quarter	Recurring General Management	Nonrecurring Technical	Recurring Technical	Recurring Structure	Recurring Outfitting	Recurring Direct Support
1st						
2nd						
3rd						
4th						
5th						
6th						
7th						
8th						
9th						
10th						
11th						
12th						
13th						
14th						
15th						
16th						
17th						
18th						
19th						
20th						
21st						
22nd						
23rd						
24th						
25th						

Unit Learning Curve (in %)

The learner curve slope is a reflection of the relative efficiency improvement (in man-hours) for repeat units. The underlying principle is that the direct labor man-hours necessary to complete a unit of production will decrease by a constant percentage each time the production quantity is doubled. If the rate of improvement is 20% between doubled quantities, then the *learning curve* would be 80% (100 − 20 = 80).

Gen. Management	Technical	Structure	Outfitting	Direct Support

Burden Rate Information

The term *burden* refers to overhead, general and administrative, and fee/profit costs. These costs are proportional to the direct hours and are, typically, billed as a percentage of the direct labor hours.

19. What burden/overhead cost pools do you use, what costs are included in each, and how are costs allocated?

20. Are there burden/overhead costs that are spread to more than one location?

21. Which do you consider fixed annual costs, and which are variable?

22. Please provide in the table below how burden/overhead changes as a function of the current business base. If you have separate burden rates for different areas/skills, please provide a rate table for each area.

% Change in Business Base	Total Direct Hours	Burden/Overhead Rate (%)	Fully Burdened (Wrap) Rate (AUD/hr)
50%			
40%			
30%			
20%			
10%			
0%			
−10%			
−20%			
−30%			
−40%			
−50%			

23. In the above table, what assumptions have you made concerning the fixed burden costs (such as asset depreciation, rent, and facilities maintenance)? Please describe.

Facilities Information

24. Given your current facilities, what do you consider your optimum workload in terms of number of programs? (Please describe in terms of numbers and types of projects.)

25. What do you consider to be the facilities that currently limit your overall capacity (e.g., dry docks, cranage, laydown areas, shops, workforce)? Please explain.

26. Do you have any planned facility upgrades, automation plans, or improvements that would increase overall throughput of the facilities? Please describe these improvements, define the timing for such upgrades, and specify how the improvements will be funded (e.g., by the company, as part of a specific modification program).

27. Please list the active, final assembly facilities (e.g., dry docks, floating dry docks, shipways, graving docks, and land-level areas) at your site(s):

#	Name	Maximum Size Ship That Can Be Accommodated			Type of Final Assembly Facility (e.g., dry dock, shipway)	How Many Ship/ Programs May Use the Facility at the Same Time? (number and types of ships)
		Length (m)	Beam (m)	Draft (m)		
1						
2						
3						

28. Please list any active outfitting piers:

#	Name	Maximum Size Ship That Can Be Accommodated		
		Length (m)	Beam (m)	Draft (m)
1				
2				

29. Please provide a rough estimate of the overall utilization of the various shops over the past year:

Shop	Approximate Percentage of Capacity Used Over Past Year	Drivers for Capacity Limit (labor, equipment, etc.)
Structure		
Piping		
Electrical		
Joinery		
Machine shop		
Paint and blast		
Other:		
Other:		

Critical Vendors

30. In this section, we ask for your help in identifying key and critical vendors, which are those who provide a critical product or service, rely almost exclusively on shipbuilding, possess a unique skill or manufacturing capability, or would likely "disappear" without some continuous demand.

31. Critical vendors:

Company	Product or Service	Value of Product or Service (AUD)	Contact	Email

In addition, what percentage of suppliers are Australian?

For the remaining non-Australian suppliers, please identify the number of suppliers by country.

Bibliography

ACIL Allen Consulting, *Naval Shipbuilding & Through Life Support, Economic Value to Australia*, ACIL Allen report to Australian Industry Group, December 2013.

Ahlgren, J., L. Christofferson, L. Jansson, and A. Linner, *Faktaboken om Gripen*, 4th ed., Linkoping, Sweden: Industrigruppen JAS AB, 1998.

Arena, Mark V., Irv Blickstein, Obaid Younossi, and Clifford A. Grammich, *Why Has the Cost of Navy Ships Risen? A Macroscopic Examination of the Trends in U.S. Naval Ship Costs over the Past Several Decades*, Santa Monica, Calif.: RAND Corporation, MG-484-NAVY, 2006. As of March 3, 2015:
http://www.rand.org/pubs/monographs/MG484.html

Arena, Mark V., Hans Pung, Cynthia R. Cook, Jefferson P. Marquis, Jessie Riposo, Gordon T. Lee, *The United Kingdom's Naval Shipbuilding Industrial Base: The Next Fifteen Years*, Santa Monica, Calif.: RAND Corporation, MG-294-MOD, 2005. As of March 16, 2015:
http://www.rand.org/pubs/monographs/MG294.html

Arena, Mark V., John F. Schank, and Megan Abbott, *The Shipbuilding and Force Structure Analysis Tool: A User's Guide*, Santa Monica, Calif.: RAND Corporation, MR-1743-NAVY, 2004. As of March 3, 2015:
http://www.rand.org/pubs/monograph_reports/MR1743.html

Australian Bureau of Statistics, "Employee Earnings and Hours, Australia," May 2013.

Australian National Audit Office, *Air Warfare Destroyer Program*, Audit Report No. 22 2013-14, March 6, 2013.

Birkler, John, Denis Rushworth, James Chiesa, Hans Pung, Mark V. Arena, and John F. Schank, *Differences Between Military and Commercial Shipbuilding: Implications for the United Kingdom's Ministry of Defence*, Santa Monica, Calif.: RAND Corporation, MG-236-MOD, 2005. As of March 2, 2015:
http://www.rand.org/pubs/monographs/MG236.html

Chang, Semoon, *Austal USA: Economic Impact on the Coastal Counties of Alabama*, Gulf Coast Center for Impact Studies, April 24, 2013.

Cockatoo Island, "Sydney's Maritime History: Ship Building," Australian government, Sydney Harbour Federation Trust, undated. As of November 21, 2014:
http://www.cockatooisland.gov.au/about/history/ship-building

Commonwealth of Australia, *2009 Defence White Paper: Defending Australia in the Asia Pacific Century—Force 2030*, Department of Defence, 2009. As of March 2, 2015:
http://www.defence.gov.au/whitepaper/2009/docs/defence_white_paper_2009.pdf

———, *2013 Defence White Paper*, Department of Defence, 2013a. As of December 8, 2014:
http://www.defence.gov.au/whitepaper/2013/docs/wp_2013_web.pdf

———, *Future Submarine Industry Skills Plan*, Department of Defence, Defence Materiel Organisation, 2013b. As of March 2, 2015:
http://www.defence.gov.au/dmo/Multimedia/FSISPWEB-9-4506.pdf

———, "Minister for Foreign Affairs and Minister for Defence—Maritime Security Strengthened Through Pacific Patrol Boat Program," Canberra, Australia: Australian Department of Defence, June 17, 2014a. As of January 15, 2015:
http://www.minister.defence.gov.au/2014/06/17/minister-for-foreign-affairs-minister-for-defence-maritime-security-strengthened-through-pacific-patrol-boat-program/

———, *Future of Australia's Naval Shipbuilding Industry: Tender Process for the Navy's New Supply Ships*, Part I, Economics References Committee, August 2014b. As of March 3, 2015:
http://www.aph.gov.au/Parliamentary_Business/Committees/Senate/Economics/Naval_shipbuilding/Report_part_1

Compass International Inc., *2014 Global Construction Costs Yearbook*, 2014.

Congressional Budget Office, *Total Quantities and Unit Procurement Cost Tables: 1974–1995*, Publication 18099, April 13, 1994. As of March 16, 2015:
http://www.cbo.gov/sites/default/files/94doc02b.pdf

Danish Defence Acquisition and Logistics Organization, "The Danish Frigate Program," presentation, November 11, 2014.

Defense Industry Daily, "Korea's New Coastal Frigates: The FFX Incheon Class," August 25, 2014. As of March 3, 2015:
http://www.defenseindustrydaily.com/ffx-koreas-new-frigates-05239/

———, "Korea's KDX-III AEGIS Destroyers," November 12, 2014. As of March 5, 2015:
http://www.defenseindustrydaily.com/drs-wins-multiplexing-contract-for-korean-aegis-destroyers-0431/

Drezner, Jeffrey, Mark V. Arena, Megan P. McKernan, Robert E. Murphy, Jessie Riposo, *Are Ships Different? Policies and Procedures for the Acquisition of Ship Programs*, Santa Monica, Calif.: RAND Corporation, MG-991-OSD/NAVY, 2011. As of March 3, 2015:
http://www.rand.org/pubs/monographs/MG991.html

Economic Development Board South Australia, *Economic Analysis of Australia's Future Submarine Program*, October 2014.

Eliasson, Gunnar, *Advanced Public Procurement as Industrial Policy*, New York: Springer, 2010.

———, "The Commercialising of Spillovers: A Case Study of Swedish Aircraft Industry," in Andreas Pyka, Derengowski Fonseca, and Maria da Graca (eds.), *Catching Up, Spillovers and Innovation Networks in a Schumpeterian Perspective*, New York: Springer, 2011, pp. 147–170.

Elmendorf, Douglas W., letter to Hon. Jeff Sessions, Washington, D.C.: Congressional Budget Office, April 28, 2010.

First Marine International Ltd., *First Marine International Findings for the Global Shipbuilding Industrial Base Benchmarking Study*, Part 2: *Mid-Tier Shipyards, Final Redacted Report*, February 6, 2007.

Garden Island Environmental Hotline, "Captain Cook Graving Dock," web page, undated. As of November 20, 2014:
http://www.gardenisland.info/1-02-010.html

GlobalSecurity.org, "Australian Shipbuilding Industry," March 27, 2012. As of November 24, 2014:
http://www.globalsecurity.org/military/world/australia/shipbuilding-history.htm

IHS, *Jane's Fighting Ships* (online), undated. As of March 3, 2015:
https://www.ihs.com/products/janes-fighting-ships.html.

Infodenfensa.com, "Special Weapons Programs Recorded a Deviation of 32% Cost," December 10, 2011, not available to the general public.

Italian Ministry of Defense, *Documento Programmatico Pluriennale per la Difesa per il Triennio 2013–2015*, April 2013. As of March 5, 2015:
http://www.difesa.it/Content/Documents/DPP_2013_2015.pdf

Japanese Bureau of Finance and Equipment, *Current Situation of Ship Production and Skill Base*, Japanese Ministry of Defense, March 2011.

Japanese Ministry of Defense, *Defense Programs and Budget of Japan: Overview of FY2010 Budget*, 2009.

———, *Defense Programs and Budget of Japan: Overview of FY2014 Budget*, December 2013.

Jones, Peter, "A Period of Change and Uncertainty," in David Stevens (ed.), *The Royal Australian Navy*, the Australian Centenary History of Defence III, South Melbourne, Victoria: Oxford University Press, 2001.

Keating, Edward G., Irina Danescu, Dan Jenkins, James Black, Robert Murphy, Deborah Peetz, and Sarah H. Bana , *The Economic Consequences of Investing in Shipbuilding: Case Studies in the United States and Sweden*, Santa Monica, Calif.: RAND Corporation, RR-1036-AUS, 2015. As of April 6, 2015:
http://www.rand.org/pubs/research_reports/RR1036.html

Lamb, Thomas, "Naval Ship Acquisition Strategies for Developing Countries," paper presented at the Pacific Northwest Section Meeting, Society of Naval Architects and Marine Engineers, Vancouver, British Columbia, November 21, 2013.

Naval Center for Cost Analysis, "Joint Inflation Calculator," spreadsheet, March 2014. As of December 8, 2014:
https://www.ncca.navy.mil/tools/inflation.cfm

Naval Sea Systems Command, *2012 Command Pocket Guide*, 2012. As of November 17, 2014:
http://www.navsea.navy.mil/OnWatch/assets/images/pocket_guide_2012.pdf

Naval Technology, "Canberra Class Landing Helicopter Docks (LHDs), Australia," web page, undated(a). As of November 24, 2015:
http://www.naval-technology.com/projects/canberra-class-landing-helicopter-docks-lhds/

———, "Juan Carlos I Landing Helicopter Dock, Spain," web page, undated(b). As of March 16, 2015:
http://www.naval-technology.com/projects/juan-carlos/

Naval Vessel Register, "Ships," web tool, undated. As of March 3, 2015:
http://www.nvr.navy.mil/nvrships/S_TYPE.HTM

OANDA Corporation, "Historical Exchange Rates," undated. As of November 29, 2014:
http://www.oanda.com/currency/historical-rates/

Odense Maritime Technology, "Australian Interest in the *Iver Huitfeldt*: Visit by RAND," presentation, November 10, 2014.

Organisation for Economic Co-operation and Development, Directorate for Science, Technology, and Industry, "Compensated Gross Ton (CGT) System," 2007.

Parliament of Australia, "Chapter 3: A Brief History of Australia's Naval Shipbuilding Industry," in *Blue Water Ships: Consolidating Past Achievements*, December 7, 2006a. As of November 18, 2014:
http://www.aph.gov.au/Parliamentary_Business/Committees/Senate/Foreign_Affairs_Defence_and_Trade/Completed_inquiries/2004-07/shipping/report/c03

———, "Chapter 4: Australian Naval Shipbuilders," in *Blue Water Ships: Consolidating Past Achievements,* December 7, 2006b. As of December 8, 2014:
http://www.aph.gov.au/Parliamentary_Business/Committees/Senate/Foreign_Affairs_Defence_and_Trade/Completed_inquiries/2004-07/shipping/report/c04

Pung, Hans, Laurence Smallman, Mark V. Arena, James G. Kallimani, Gordon T. Lee, Samir Puri, and John F. Schank, *Sustaining Key Skills in the UK Naval Industry*, Santa Monica, Calif.: RAND Corporation, MG-725-MOD, 2008. As of March 16, 2015:
http://www.rand.org/pubs/monographs/MG725.html

Reiner, Daniel, Xavier Pintat, and Jacques Gautier, *Défense: Équipement Des Forces et Excellence Technologique Des Industries De Défense*, Senate Presentation, November 21, 2013. As of March 5, 2015:
http://www.senat.fr/rap/a13-158-8/a13-158-81.pdf

Roos, Goran, "Future of Australia's Naval Shipbuilding Industry," supplementary submission to the Senate Economics References Committee, October 13, 2014.

Royal Australian Navy, "*HMAS Rankin*," web page, undated.

———, "The Pacific Patrol Boat Project, *Semaphore: Newsletter of the Sea Power Centre—Australia*, No. 2, Canberra, Australia: Department of Defence, February 2005. As of December 8, 2014:
http://www.navy.gov.au/media-room/publications/semaphore-february-2005

Schank, John F., Mark V. Arena, Paul DeLuca, Jessie Riposo, Kimberley Curry, Todd Weeks, James Chiesa, *Sustaining Nuclear Submarine Design Capabilities*, Santa Monica, Calif.: RAND Corporation, MG-608-NAVY, 2007. As of November 3, 2014:
http://www.rand.org/pubs/monographs/MG608.html

Schank, John F., Mark V. Arena, Denis Rushworth, John Birkler, and James Chiesa, *Refueling and Complex Overhaul of the USS Nimitz (CVN 68): Lessons for the Future*, Santa Monica, Calif.: RAND Corporation, MR-1632-NAVY, 2002. As of November 3, 2014:
http://www.rand.org/pubs/monograph_reports/MR1632.html

Schank, John F., Cynthia R. Cook, Robert Murphy, James Chiesa, Hans Pung, and John Birkler, *The United Kingdom's Nuclear Submarine Industrial Base*, Vol. 2: *Ministry of Defence Roles and Required Technical Resources*, Santa Monica, Calif.: RAND Corporation, MG-326/2-MOD, 2005. As of March 16, 2015:
http://www.rand.org/pubs/monographs/MG326z2.html

Schank, John F., Cesse Ip, Francis W. LaCroix, Robert E. Murphy, Mark V. Arena, Kristy N. Kamarck, and Gordon T. Lee, *Learning from Experience*, Vol. 2: *Lessons from the U.S. Navy's* Ohio*, Seawolf, and* Virginia *Submarine Programs*, Santa Monica, Calif.: RAND Corporation, MG-1128/2-NAVY, 2011. As of November 3, 2014:
http://www.rand.org/pubs/monographs/MG1128z2.html

Schank, John F., Jessie Riposo, John Birkler, and James Chiesa, *The United Kingdom's Nuclear Submarine Industrial Base*, Vol. 1: *Sustaining Design and Production Resources*, Santa Monica, Calif.: RAND Corporation, MG-326/1-MOD, 2005. As of March 16, 2015:
http://www.rand.org/pubs/monographs/MG326z1.html

Smallman, Laurence, Hanlin Tang, John F. Schank, and Stephanie Pezard, *Shared Modular Build of Warships: How a Shared Build Can Support Future Shipbuilding*, Santa Monica, Calif.: RAND Corporation, TR-852-NAVY, 2011. As of March 9, 2015:
http://www.rand.org/pubs/technical_reports/TR852.html

Spanish Ministry of Defense, *Evaluación de los Programas Especiales de Armamento (PEAs)*, Madrid: Grupo Atenea, September 2011.

Stevens, David, *The R.A.N.—A Brief History*, website, Royal Australian Navy, undated. As of November 18, 2014:
http://www.navy.gov.au/history/feature-histories/ran-brief-history

Straczek, J. H., "RAN in the Second World War," website, Royal Australian Navy, undated. As of November 18, 2014:
http://www.navy.gov.au/history/feature-histories/ran-second-world-war

Thornton, Sean, "The Rationale for the RAN Offshore Combatant Vessel," *The Navy (Navy League of Australia)*, Vol. 72, No. 1, January 2010, pp. 6–10.

Todd, Daniel, and Michael Lindberg, *Navies and Shipbuilding Industries: The Strained Symbiosis*, Westport, Conn.: Praeger Publishers, 1996.

UK National Audit Office, *The Major Projects Report 2000*, U.K. Ministry of Defence, November 22, 2000.

———, *The Major Projects Report 2011*, U.K. Ministry of Defence, November 16, 2011.

UK Office for National Statistics, "Weekly Pay—Gross (£)—For Full-Time Employee Jobs: United Kingdom, SIC2007, Table 16.1a," 2013.

U.S. Bureau of Labor Statistics, "International Comparisons of Hourly Compensation Costs in Manufacturing," May 2013a.

———, "National Industry-Specific Occupational Employment and Wage Estimates: NAICS 336600—Ship and Boat Building," May 2013b.

U.S. Department of Defense, *Cost Analysis Guidance and Procedures*, DoD 5000.4-M, Washington, D.C., December 11, 1992.

U.S. Department of the Navy, *Department of the Navy Fiscal Year (FY) 2013 Budget Estimates: Justification of Estimates, Shipbuilding and Conversion, Navy*, February 2012.

———, *Department of the Navy Fiscal Year (FY) 2015 Budget Estimates: Justification of Estimates—Shipbuilding and Conversion, Navy*, March 2014.

Willett, Andrew, *Air Warfare Destroyer (DDG) Future Personnel and Operating Cost (FPOC) Report 2014*, Version 5.2, October 2014.

Yule, Peter, and Derek Woolner, *The Collins Class Submarine Story: Steel, Spies and Spin*, Cambridge, U.K.: Cambridge University Press, 2008.